AFRICAN ENVIRONMENTAL

and

HUMAN SECURITY

in the 21st Century

AFRICAN ENVIRONMENTAL
and
HUMAN SECURITY
in the 21st Century

edited by
HELEN E. PURKITT

CAMBRIA
PRESS

Amherst, New York

Requests for permission should be directed to:
permissions@cambriapress.com, or mailed to:
Cambria Press
20 Northpointe Parkway, Suite 188
Amherst, NY 14228

Chapter 1 contains a slightly revised version of "Human and Environmental Security in the Sahel: A Modest Strategy for Success," in *Environmental Change and Human Security: Recognizing and Acting on Hazard Impacts*, edited by P. H. Liotta, David A. Mouat, William G. Kepner, and Judith M. Lancaster, 341–393 (Dordrecht: The Netherlands, 2008). This publication is part of the NATO Science for Peace and Security Series C: Environmental Security, and is reprinted with the permission of the editor and Springer. The original publication is available at http://www.springer.com. Chapter 2 first appeared as Stephen F. Burgess, "Environment and Human Security in the Horn of Africa," *Journal of Human Security* 4, no. 2 (2008): 37–61. The article is reprinted with the permission of the publisher of *Journal of Human Security*.

Library of Congress Cataloging-in-Publication Data

African environmental and human security in the 21st century / edited by Helen E. Purkitt.
        p. cm.
    Includes bibliographical references and index.
    ISBN 978-1-60497-646-5 (alk. paper)
1. Africa—Environmental conditions—21st century. 2. National security—Environmental aspects—Africa—History—21st century. I. Purkitt, Helen E., 1950– II. Title.

    GE160.A35A355 2010
    363.70096'0905—dc22

2009038312

To all worldwide who are working to implement
a more comprehensive form of security
in their immediate, regional, and global environments

TABLE OF CONTENTS

LIST OF FIGURES AND TABLE

PREFACE

The volume evolved from a set of papers presented at a panel entitled Environmental and Human Security in Africa during the 49th International Studies Association (ISA) Annual Convention in March 2008. The purpose of the panel was to identify more systematic ways to understand, monitor, and remediate specific human and environmental problems in Africa. A common theme among panelists was the importance of understanding fundamental causal mechanisms so that effective policies and actions can prevent human and environmental problems from worsening in the future.

Since all of the panel paper presenters worked for U.S. military educational, research, policy, or operational organizations, there was a shared interest in identifying ways that U.S. organizations, including the recently established AFRICOM, can provide effective, nontraditional security aid. Each paper identified specific measures where remarkably modest amounts of aid or assistance on the part of U.S. or other outside actors might be extremely useful in one or more environmental or human security issue areas. The papers focused on such diverse problems as how to establish security arrangements in transfrontier nature parks (chapter 10), enhance human security in Niger and Chad (chapter 1), understand and improve early warning systems and interregional cooperation for coping with droughts and other humanitarian emergencies in the Horn and East Africa (chapter 2).

Dr. Geoffrey D. Dabelko, director, Environmental Change and Security Program, Woodrow Wilson International Center for Scholars, Washington, DC, and retired colonel Maxie McFarland, the senior Africa analyst and deputy chief of staff for Army Intelligence, served as discussants and offered comments and critiques of the papers. During the panel, Colonel McFarland noted that the U.S. military was already doing a lot of nontraditional, human security-type projects in Iraq and Afghanistan, but cautioned that just because the army (and other branches of the military) can do this type of activity, "doesn't mean you want them to do it." This cautionary comment echoed caveats found in several of the chapters in this volume regarding recommended future activities by AFRICOM, USAID, and other U.S. agencies. Maxie McFarland expands on his earlier comments in the concluding chapter of the volume (chapter 11).

The volume also includes several additional studies that provide indepth case studies of specific conflicts or problem areas that have the potential to become larger African human and environmental issues in the future. Two of the studies use new interdisciplinary approaches to analyze and monitor infectious diseases (chapter 5) and public policies designed to manage problems created by climate change (chapter 6). Three chapters are policy analyses related to terrorism and the environment (chapter 7), the illegal exploitation of natural resources (chapter 8), and peacekeeping operations and the environment (chapter 9).

The organization of the volume reflects our shared view that an adequate understanding of complex human and environmental security problems requires high quality, case-specific knowledge, the application of multidisciplinary and multilevel conceptual approaches, and the ability to develop policy-relevant, practical recommendations from diverse bodies of knowledge. Part 1, "Human and Environmental Conflicts: Conceptualization and Case Studies," contains Colonel Jebb, U.S. Army, et al.'s in-depth comparative case studies of Niger and Chad using a multilevel and inter-disciplinary systems approach. In chapter 2, Associate Professor Steven Burgess examines the linkages among ethnic conflict and four different types of environmental security threats in the Horn and East Africa. He relates these conflicts to well-known theories of how environmental deg-radation and scarcity are linked to political conflicts. U.S. Marine Major Mark Deets, a former defense attaché in Senegal, the Gambia, and Guinea-Bissau, analyzes the root causes of the continuing low-level conflict in the Casamance region of Senegal. Major Deets found that desertification was one of several drivers of continued low-level conflict. He concludes that unless there are major changes in the region, such as the increased use of the Casamance region as a transit area for smuggling people, arms, and drugs, the conflict is likely to continue at a low level of intensity for the foreseeable future.

The last in-depth case study in this section (chapter 4) is by Dr. Anthony Turton, a well-known expert on water issues in southern Africa. Dr. Turton discusses the factors and empirical evidence in South Africa that led him to hypothesize that "when the expectations of society for resource allocation exceeds the capacity (or willingness) of the government to deliver, then mass violence could result." In his analysis, Dr. Turton notes that water is already the single most important resource that constrains future develop-ment potential in the country. He ties the modern struggle over water to the country's brutal past struggles to control and allocate valuable resources, such as water and mineral wealth, before and during the apartheid era, and to the breakdown of water sanitary procedures since 1994, including increased evidence that decanted water from mines now contains radioac-tive and heavy metal concentrations that are flowing into rivers and into industrial and home water supplies and may already be causing adverse human health effects.[1] Dr. Turton had planned to discuss the need for more research on the harmful effects of growing water pollution in South Africa in a speech he was invited to make at a plenary session of South Africa's Center for Science and Industrial Research (CSIR) in November 2008. Shortly before his presentation, Dr. Turton was prevented from delivering the speech by his superiors and was later released from his employment at CSIR. Although his speech was banned, it was immediately posted on the Internet by a nongovernmental organization and remains available to interested members of the public.[2] In his analysis for this volume, Dr. Turton explains his reasoning for predicting that if current water pollution trends are allowed to continue unchecked, South Africa is likely to experience more adverse environmental and adverse health effects, in addition to future mass political violence. Dr. Turton proposed his hypothesis in the hopes that researchers in the wider international community, after reviewing the available evidence, will agree that much more research in this area is needed.

Part 2, "Emerging Transcontinental Issues: Conceptualizations and Case Studies," covers new ways to monitor human and environmental problems and how government and other organizations are attempting to manage these problems. Chapter 5 develops a framework based on concepts drawn from epidemiology and political risk assessments to understand the different types of costs and risks associated with neglecting spreading diseases, including the spread of Guinea worm disease in conflict areas throughout Sudan and cholera in the state of Zimbabwe. Dr. Chad Briggs and Dr. Jennifer Bath's framework looks promising as a method for monitoring the multidimensional or second- and third-order effects of systemic collapse in the community and wider area due to unchecked spread of diseases. Chapter 6 provides a multidimensional framework and methodology for assessing future risks and evaluating the ability of African governments to cope with environmental problems within six policy categories, using ten performance indicators designed to measure the environmental burden of disease, water, and air pollution, biodiversity and habitat, forestry, fisheries, agriculture, and climate change. The author, Assistant Professor John Ackerman of the National Security Studies at the Air Command and Staff College, compares standardized national performance indicator scores for Egypt, Nigeria, South Africa, and Kenya with those of economic and regional peer countries. These comparisons are offered as one approach for assessing the performance of national governments in different policy areas and making recommendations for future actions. The last chapter in this section is by a practitioner turned academic. Colonel Kent Butts (retired) provides an overview of environmental security issues in Africa and how these environmental issues figure into U.S. efforts to counter jihadist terrorism on the continent. Colonel Butts discusses how cooperation over environmental and human security issues has often formed the basis for long-lasting relationships among parties who often have very different definitions and views as to priorities and meanings of "security" and "terrorism" in African countries.

The final section of this volume offers recommendations, remediation, and some concluding thoughts. In chapter 8, Elisabeth Feleke, regional program manager of the newly formed West Africa office of the U.S. government agency, the Africa Center for Strategic Studies (ACCS), discusses several ongoing African conflicts where "greed and grievances" by locals fuel the exploitation of natural resources and violent conflicts. Feleke also offers several recommendations about what outside agencies can do to help resolve or minimize the exploitation of natural resources by participants in African conflicts. In chapter 9, Lt. Col. Brent C. Bankus [retired], affiliated with the Operations and Gaming Division of the Center for Strategic Leadership at the U.S. Army War College, uses examples of recent exploitation of natural resources by both combatants and some peacekeepers in his discussion of the ways that some key international organizations are involved in efforts to prevent the exploitation of people and natural resources in Africa. In chapter 10, Dr. Dan Henk, director of the Air Force Culture and Language Center at the Air War College, provides a useful overview of major trends in efforts to establish African transfrontier conservation areas before turning to a detailed analysis of the impetus, status, and problems associated with implementing a plan for the single largest "peace park" in Africa, the Kavango-Zambezi, or KAZA. Dr. Henk notes that this project, if implemented successfully, will promote human and environmental security. However, Dr. Henk also cautions that participating nation-states must be able to work out mutually agreeable mechanisms for maintaining security in the nature areas and sharing future revenues if the project is to succeed. He concludes by offering suggestions about how the United States and other outside actors may facilitate the planning and implementation of acceptable security mechanisms for the planned nature area, which will encompass large amounts of territory in four African nation-states. Colonel Maxie McFarland's essay provides some concluding remarks about the findings of the original set of papers and offers some recommendations about the future role that AFRICOM, and the U.S. military more generally, should play in African environmental security issues.

Collectively, these works represent an impressive body of theory, analysis, and practical advice that should be of interest to a wide audience, including civilian and military practitioners who are working or plan to work in Africa or other parts of the developing world; students of African politics, security, and environmental or developmental studies; and concerned citizens worldwide. As we approach the real limits of growth on our planet, the questions addressed in this volume related to promoting human and environmental security in Africa are likely to become relevant to an even larger audience, especially once it is recognized that the security interests of those living in the developed world are inextricably linked to the security interests of Africans and those living in other parts of the world.

ACKNOWLEDGMENTS

I would like to thank all of the contributors for their thoughtful essays and analyses. I would also like to thank Dr. Paul Richardson for his initial interest in the project and continuing encouragement and Toni Tan, director at Cambria, for her support of the project and patience in helping to complete the manuscript. A special thanks goes to the late Robert Rivera for his work on the final manuscript and to Dr. Julie Zhu for her technical assistance in finalizing the image files.

AFRICAN ENVIRONMENTAL

and

HUMAN SECURITY

in the 21st Century

PART I

HUMAN AND ENVIRONMENTAL CONFLICTS

CONCEPTUALIZATION AND CASE STUDIES

CHAPTER 1

ON THE MARGINS

INSECURITIES IN NIGER AND CHAD[3]

*Cindy R. Jebb, Laurel J. Hummel,
Luis Rios, and Madelfia A. Abb* [4]

> No man is an island, entire of itself; every man is a piece of the
> continent, a part of the main. If a clod be washed away by the sea,
> Europe is the less, as well as if a promontory were, as well as if
> a manor of thy friend's or of thine own were: any man's death
> diminishes me, because I am involved in mankind, and therefore
> never send to know for whom the bell tolls; it tolls for thee.
> —John Donne, "Meditation XVII"
> of *Devotions upon Emergent Occasions* (1623)

INTRODUCTION

John Donne's 1623 passage profoundly recognizes the interconnected-
ness of humanity and the living world. If a bell were to ring for every

human being who dies due to human insecurities, the bell would never cease to ring. In the case of many African states, the bell rings constantly. Paraphrasing Donne, every African life that ends diminishes all of us in many and various ways. What happens in Africa affects the rest of the globe, including the most prosperous and technologically advanced states. The global effects of human insecurities in Africa are real, because everything is connected and related. The current War on Terrorism elicits a more narrow view of these connections by focusing on just an outward manifestation of deteriorating security environments. To truly understand the challenges of terrorism perpetrated by radical militants and extremists, we must understand the totality of the security environment while understanding its unique nature from place to place.

We use a human security approach to accomplish these tasks, that is, understanding the totality and uniqueness of a particular security environment. This chapter's focus on the roots of conflict and instability in the Sahel is timely and critical as the United States establishes a new African command and the work of the Trans-Sahel Counterterrorism Partnership (TSCTP) continues.[5] We use primarily the case of Niger and a secondary case, Chad, to highlight the importance of understanding security issues in a comprehensive and holistic manner, with added emphasis on understanding the environmental impact on human survival and quality of life. However, we acknowledge that there is no one-size-fits-all policy; instead, successful policy requires tireless, smart, and patient work. Specifically, the authors—a political scientist, a biologist, a human geographer, and a climatologist—will highlight the strong connection between human survival and the environment. We acknowledge that, unfortunately, there have been numerous interpretations and definitions of human security, so it has come to mean everything to everyone. For example, when security professionals of powerful states use this term, they arouse suspicions of cloaking interventions and neocolonialism in the name of protection. We contend, however, that the human security paradigm fosters a holistic and empathetic approach towards understanding the security environment. The chapter uses on-the-ground experience and interviews, political analysis, and geographical analysis to understand how the environment cuts across all areas of human security, as explained by the UNDP and measured by the 2000 Millennium Development Goals discussed below.

This holistic approach is not only critical for understanding the security environment, it is essential for strategists to better anticipate second- and third-order effects of their decisions and subsequent actions. Moreover, it puts the fight against terrorism in its proper context so that tactical counterterrorist actions do not undermine the strategic goals of mitigating root causes that, in the long term, will help "drain the swamp" of potential recruits. As a first step, we will briefly explain why human security matters by viewing the security environment through the lens of a biologist, using the living systems theory. Second, we will explain human security as an approach and operational concept. Third, we will emphasize the importance of the environment as it relates to all areas of human security. Our case studies on Niger and Chad from the Sahel region will illuminate these points before we conclude with environmental and policy implications.

WHY HUMAN SECURITY MUST BE EXPLORED

In short, human security describes threats that states cannot or will not deter, that directly affect individuals. To fully appreciate this paradigm, however, requires a "shift of mind."[6] The living systems theory (LST) drives this shift with its biological view of the world. This perspective sheds light on the nonlinear nature of actions and behaviors. In other words, actions are not constricted to the parameters of cause-and-effect relationships. Instead, due to myriad connections and relationships, events and activities have second- and third-order effects that are not always recognizable. Moreover, there are various unintended consequences of even the best-planned events. The best we can do is study behavior from multiple perspectives to gain a holistic and empathetic understanding. According to Fritjof Capra, "The behavior of a living [system] organism as an integrated whole cannot be understood from the study of its parts alone."[7] For the whole is more than the sum of its parts.[8] Robert Jervis explains that the whole is indeed different from the sum of its parts.[9] Through the lens of LST, the world is a living system consisting of multiple systems within systems, with an almost infinite number of relationships and connections among all systems, living and nonliving.[10]

The foundation of the living systems theory is from the theoretical works of James G. Miller and Fritjof Capra. In his book *Living Systems*, Miller proposes that complex structures which carry out living processes can be identified at seven hierarchical levels and have nineteen critical subsystems whose processes are essential for life.[11] Types of processes that are essential for life are the processing of matter, energy, and information. According to Miller, there are seven hierarchical levels of living systems: the cell, organ, organism, group, organization, society, and supranational. The hierarchy of living systems begins at the simplest living system (the biological cell) and ends at the most complex living system (the supranational system, which consists of two or more societal systems). Miller painstakingly details the increasing complexities of his nineteen processes for each level of living systems.

Fritjof Capra offers another approach, which we adopt. In his book *The Web of Life: A New Scientific Understanding of Living Systems*, Capra theorizes a living system possesses form, matter, process, and meaning.[12] These four criteria of living coexist and enable living systems to adapt, survive, create, and propagate through time. These characteristics reveal that living systems make choices and have unique worldviews. Examples of living systems include societies, governments, terrorist organizations, and other human groupings, as well as environmental elements such as animals, crops, trees, and so on.[13] These living systems also interact with nonliving systems such as the soil, air, and water. These insights are important for three reasons: First, they point to the importance of understanding the impact of the security environment, in all its manifestations, on the individual and how that individual decides to interact with other living and nonliving systems around him or her. It also reminds us that groupings of individuals make choices as well: that is, governments, societies, ethnic groups, terrorist organizations, and so on. Second, there are myriad connections and relationships among and between these systems. We cannot possibly explore all possible relationships among living and nonliving systems, but we can focus on those relationships between political organizations, societies, and terrorist organizations that are vying for legitimacy. This point leads to our final concern, reflected in the criterion of "meaning."

This fourth criterion, meaning, speaks to the human and social domains of living systems.[14] It reflects the rules of behavior, values, intentions, goals, strategies, designs, and power relations essential to human social life.[15] In essence, "meaning" refers to the unique perspective or view of the world a system has, based on its unique qualities. For the purposes of this discussion, this criterion refers to a system's mental map. For example, while the United States may regard Al-Qaeda members as terrorists, Al-Qaeda members view themselves as martyrs. Different societies—based on their unique historical experiences, culture, and norms—have different perspectives on the world and view world events differently. This worldview shapes, for example, the quality of the relationship individuals or groups of individuals have towards a particular state, and legitimacy measures the quality of those state-societal connections. This is a key point as states work to offer their people security in a bid to gain legitimacy vis-à-vis extremists and terrorists. Francis Fukuyama notes the importance of state legitimacy: "The state's institutions not only have to work together properly as a whole in an administrative sense, they also have to be perceived as being legitimate by the underlying society."[16] Security, however, must be interpreted from the individual's perspective because it is the individual who ultimately decides on his or her behavior and whether or not the state has earned his or her loyalty. Subsequently, we now explore human security as a means to gain this holistic and empathetic perspective.

HUMAN SECURITY[17]

When the UN presents aggregate data such as 1 billion people who lack access to clean water, 2 billion people who lack access to clean sanitation, 3 million people who die from water-related diseases annually, 14 million people (including 6 million children) who die from hunger annually, and 30 million people in Africa alone who die of HIV/AIDS, it primarily describes the myriad human insecurities of underdeveloped regions and the increasing global gap between the haves and the have-nots.[18] Many of the human insecurities exist in the non-Western world, primarily in underdeveloped regions. These regions consist of weak states with porous borders, overlapping ethnicities, and colonial histories. Many of these weak states entered the international system late and, interestingly, attained legitimacy from the international system but not from their own citizenry. Moreover, the threats facing these states have historically been internal, many times reflecting ineffective institutions, a weak state identity, a lack of state capacity, and a bankrupt economy, as well as dangerous neighbors. Clearly, the formation of many of these states did not follow the Western, Westphalian path. Some scholars suggest that their development has led to state-nations as opposed to nation-states.[19]

While there are some commonalities among underdeveloped regions, they are far from all alike. One commonality, however, is that many of these transnational forces are directly harmful to people, and often the state either cannot or will not deter or address them. As the name implies, these transnational threats do not recognize borders and are particularly harmful to susceptible regions that are already weak in myriad ways. Moreover, many of these forces are interrelated and reinforcing, and therefore exacerbate conditions that create widespread suffering.

Human security is a concept that both describes these conditions and provides an approach to better understand their effects. Due to variations among regions and localities, a first step in designing strategy is to understand the dynamics and circumstances at the ground level. The UN recognized this approach in 1994:

> The concept of security has far too long been interpreted narrowly: as security of territory from external aggression, or as protection of national interests in foreign policy or as global security from the threat of nuclear holocaust. It has been related to nation-states more than people...Forgotten were the legitimate concerns

of ordinary people who sought security in their daily lives. For many of them, security symbolized protection from the threat of disease, hunger, unemployment, crime [or terrorism], social conflict, political repression and environmental hazards. With the dark shadows of the Cold War receding, one can see that many conflicts are within nations rather than between nations.[20]

In 2004 the European Union's Human Security Doctrine for Europe used human security as its strategic anchor, defining it as "individual freedom from basic insecurities...genocide, widespread or systematic torture and degrading treatment, disappearances, slavery, and crimes against humanity and grave violations of the laws of war as defined in the Statute of the International Criminal Court."[21]

Even the UN has changed the definition of sovereignty to adjust to today's security environment:

> In signing the Charter of the United Nations, States not only benefit from the privileges of sovereignty but also accept its responsibilities. Whatever perceptions may have prevailed when the Westphalian system first gave rise to the notion of State sovereignty, today it clearly carries with it the obligation of a State to protect the welfare of its own peoples and meet the obligations to the wider international community.[22]

Consequently, when a state cannot or will not protect its citizens, then the international community must engage.[23] Scholars have described this forward-leaning definition of international security as the "duty to protect."[24] While human insecurities primarily occur in weak and failing states, due to the nature of weak regions, such insecurities quickly spread. And, as we have witnessed in our increasingly global world, these insecurities frequently have diffuse global effects, such as migrations, reverberations in diaspora communities, environmental impacts, and even the exportation of terrorism.[25] The human-security paradigm should remind strategists that they must approach issues holistically and empathetically. This requires painstaking analysis, patience, and tenacity.

OPERATIONALIZING HUMAN SECURITY:
UNDERSTANDING ITS TRANSNATIONAL COMPONENTS

The UNDP report identified components of human security. They are:

- Economic security—the threat is poverty.
- Food security—the threat is hunger and famine.
- Health security—the threat is injury and disease.
- Environmental security—the threat is pollution, environmental degradation, and resource depletion.
- Personal security—the threat involves various forms of violence.
- Community security—the threat is to the integrity of cultures.
- Political security—the threat is political repression.[26]

The 2000 Millennium Development Goals (MDGs) provide a measure towards reducing human insecurities. They are:

- Goal 1: Eradicate extreme hunger and poverty.
- Goal 2: Achieve universal primary education.
- Goal 3: Promote gender equality and empower women.
- Goal 4: Reduce child mortality.
- Goal 5: Improve maternal mortality.
- Goal 6: Combat HIV/AIDS, malaria, and other diseases.
- Goal 7: Ensure environmental stability.
- Goal 8: Develop a global partnership for development.[27]

Taken together, most areas suffering from human insecurities are facing issues of economic underdevelopment, environmental degradation and resource scarcity, food insecurity, health insecurities, political and/or civil inequalities, and violence, including human trafficking, terrorism, crime, and armed conflict. Many insecure areas also are vulnerable to radical ideologies that stifle development and exacerbate social tensions. Clearly, not all areas face the same insecurities or combination of insecurities, and, therefore, there is no common strategy for all situations. What strategists can count on is some combination, interconnection, and regional effect of these issues. With this in mind, the following is a brief description of these issues.

Economic Insecurities

The first MDG is to eradicate hunger and poverty. How does one measure poverty? How meaningful is the "one-dollar-a-day" measurement of basic subsistence requirements, given that this figure does not capture the full extent of misery that people experience? The dollar-a-day measurement presumes that X amount of capital brings equal access to commensurate commodities, which is a spurious comparison. If that amount does not allow people to have the basic necessities of life—food, clean water, health, and so on—then perhaps this measure lacks meaning.[28] The UNDP claims that the best measure of global inequality is the global income distribution model. This model uses national household income data to build an integrated global income distribution model. This model suggests a huge gap between the very rich and very poor on a global scale that is greater than the inequalities within any one country. Two-thirds of this gap is caused by income inequalities among countries, while inequalities within countries are responsible for the other third.[29] The conditions of extreme poverty make clear their immediate connection to "disease, drought, and distance from world markets."[30] Jeffrey Sachs, the world's leading economist, commented, "I began to suspect that the omnipresence of disease and death had played a deep role in Africa's prolonged inability to develop economically."[31] Further, he stated that "good governance and market reforms are not sufficient to guarantee growth if the country is in a poverty trap."[32]

Unfortunately, globalization has indeed created a poverty trap and hardened the distinctions between the haves and have-nots. Though fewer people actually live below the poverty line of one U.S. dollar per day, inequalities have, in fact, grown. According to the UNDP, stagnation has been a prevalent aspect of globalization, notably within twenty-five sub-Saharan African countries and ten Latin American countries. In Russia, after the Asian financial crisis, 30 million people slipped below the poverty line, and in Argentina between 2000 and 2003, the number of people below the poverty line tripled. While incomes have been rising in China and India and contributing, therefore, to a more positive aggregate picture of poverty reduction, this has masked the deep inequalities that exist across regions.[33] Such chronic poverty corresponds to human misery, as Jeffrey Sachs noted earlier. When faced with such a situation, people look for relief, whether through migration, rampant urbanization, or other nonstate actors. These patterns and social dislocations stemming from economic insecurities may affect social, political, and economic development not only within a particular state but also within the larger region.

Food Insecurities

While recent estimates claim that about 1 billion people are undernourished, including over 800 million in the developing world, most projections indicate that by 2050, world population will approach 9 to 10 billion.[34] It is not a coincidence that the first MDG lists the eradication of both hunger and poverty, as they are intricately related. This relationship is increasingly apparent in developing countries, rural areas, and growing urban centers.[35] Food security is at the heart of every aspect of human life. Many cultures, notably in the developing world, are centered around food-getting strategies. Though Thomas Malthus in the late eighteenth century warned of the consequences of an unchecked, exponentially growing population and a linear increase in food supplies, there may be other factors that require study that were not evident, or at least not considered, in the eighteenth and nineteenth centuries.[36] For example, food shortages may also be caused by prevailing societal, political, and economic structural sources that artificially and adversely affect food supplies.[37] Henrique Cavalcanti describes three pillars of food security: availability, access, and stability. He defines food security as the

> practices and measures related to the assurance of a regular supply and adequate stocks of foodstuffs of guaranteed quality and nutritional value…broadly defined, food security can only be achieved if every person, family, community and nation is considered in the process and at the same time play a responsible role, as sustainable consumers, in ensuring its efficiency and effectiveness.[38]

Policy makers must also consider the regional effects of food security. For example, during the 2005 famine (or what others referred to as the food crisis) in Niger, it was clear that food aid was entering the country only to be drifting south to Nigeria. Unfortunately, when Nigeria resold the food to Niger, the Nigeriens suffering from hunger could not afford to purchase the food that was originally intended for them. Food insecurities know no boundaries, especially among countries that share growing environmental depletion of food sources.

One of the factors that Malthus could not forecast is technology. Between 1950 and 1984, grain production outpaced population increase; however, after 1984, grain harvests declined, especially in Africa. Some of the major causes included soil erosion, desertification, transfer of cropland to nonfarm purposes, falling water tables, and rising temperatures. Moreover, shocks such as natural disasters, droughts, and disease can contribute to food insecurities. These conditions may reveal themselves globally in the form of higher food prices, which further emphasize the disparity between the haves and the have-nots, but at the state level, these conditions can develop into grave instabilities and conflict.[39]

Health Insecurities

It should be no surprise that the prevalence of HIV/AIDS in Africa directly exacerbates food insecurities. Clearly, the loss of workers due to AIDS adversely affects the ability to harvest crops, for example.[40] In some African countries, HIV/AIDS has infected 20 to 30 percent of the adult population.[41] But Africa is not alone. According to Nicholas Eberstadt, "major epidemics are already underway in China, India, and Russia, and the local social mores and behavioral practices are set to further spread the disease."[42] Of course, HIV/AIDS is but one aspect of health insecurity. On a recent trip to Africa, the authors were appalled by the prevalence of malaria. Water-related diseases account for 3.4 million deaths per year, and 1 billion people lack access to clean water. Even in the United States, 218 million people live within ten miles of a polluted body of water—and that is more than thirty years since the legislation of the Clean Water Act.[43] The recent scare of bird flu and SARS reveals how quickly disease can spread and highlights the need for a responsive, preventative, and proactive public health policy.

The connection between the rise of infectious diseases and the degradation of the environment is clear. According to the World Health Organization (WHO),

> poor environmental health quality is directly responsible for some 25 percent of all preventable ill health, with diarrheal diseases and acute respiratory infections heading the list. Two-thirds of all preventable ill health due to environmental conditions occurs among children, particularly the increase in asthma. Air pollution is a major contributor to a number of diseases and to a lowering of the quality of life in general.[44]

Moreover, the 2005 *Millennium Ecosystem Assessment* report claims that continued degradation of ecosystems may increase the spread of common diseases such as malaria, as well as facilitate the evolution of new diseases.[45] States with inadequate (or no) public health support systems will be most vulnerable and least able to respond. Even developed states that do have such systems will be stressed, and if faced with a true health emergency, may be ineffective, possibly losing society's trust. Of course, food and environmental insecurities and poverty exacerbate health insecurities, and vice versa. While developed countries are facing declining birth rates and the health challenges associated with aging populations, developing countries' health insecurities, notably high infant and child mortality rates, help to drive high birth rates, as parents want to ensure that at least some of their children reach adulthood. Ironically, it is these areas that suffer the most acute array of human insecurities that have the highest population growth rates, a combination of factors that exponentially increases misery and suffering.

Radicalization
Bruce Hoffman explained that terrorism is dangerous, but the real danger is the potential it has to spark a political movement. It is these radical ideas that pose a great threat to peace and stability.[46] Why is radical Islam

such a threat? First, with the end of the Cold War, democracy's ideological competitor, communism, was defeated. This radical form of political Islam has filled an ideological vacuum. It is dangerous because it appeals to people who are suffering from societal, political, and economic despair and seemingly have nowhere else to turn. Unfortunately, this radicalization prohibits long-term development as measured by the components of human security. For example, in the southern areas of Niger that are influenced by Nigeria's northern Islamic radical elements, women are told not to vaccinate their children because the vaccinations are a conspiracy of the West to sterilize them. Consequently, children are contracting preventable diseases such as polio. The 2002 UNDP *Arab Human Development Report* noted the three deficits facing the Arab world: "the freedom deficit; women's empowerment deficit; and, the human capabilities/knowledge deficit relative to income."[47] A radical agenda does not address these deficits for the Arab region or any other region and leaves the most marginalized segment of the population in status quo.

Even the radicals themselves put aside ideology for pragmatism. They are able to identify vulnerable groups and develop loyalties in place of states that cannot deliver basic needs to their population. Herein lies the danger of weak and failing states. Not only do they provide safe havens for terrorists, either wittingly or unwittingly, but they also present extremely vulnerable populations that are susceptible to radicalization in the absence of any alternatives.

Political and Civil Inequalities and Violence
Conditions that magnify a population's vulnerabilities stem from illegitimate governments that stifle political and civil equalities while failing to protect and enforce rule of law. An inability to enforce rule of law invites a potent mix of criminals, terrorists, smugglers, human traffickers, and so on, who have a tendency to reinforce each other's activities. These various criminal and terrorist groups have developed alliances due to both globalization and the availability of weak and failing states. "The absence of rule of law...provides ideal conditions for the blending of criminal and terrorist activities. Economic hardship in many of these

nations leads to corruption and trafficking of illicit goods…"[48] One type of good that has greatly affected the rise of violence is the accessibility of small and light arms that either finance or are used in terrorist or criminal activities.[49] And, unfortunately, there still is trafficking in human beings, though, for example, Niger in 2004 made slavery a criminal offense. However, due to the government's inability to enforce the law, there are a reported 43,000 slaves in Niger.[50] According to Reven Paz and Moshe Terdman, Africa offers opportunities for terrorists:

> The political and military conditions in most of the African continent, the broad weakness of its governments, and the internal fighting and corruption of these regimes, ease the ability of the Mujahidin to move, plan, and organize themselves, far from being seen. They enjoy in Africa easier operational abilities than in other countries, which have effective security, intelligence, and military capacities.[51]

The illegitimate and poorly functioning regimes provides a set of conditions that invites a potent mix of violent groups. This only serves to exacerbate human insecurities.

Environmental Insecurities
We argue that environment insecurities have profound effects on all the other human security components. For example, according to Lester Brown, economic development will be meaningless if we do not have a planet that can sustain a global economy.[52] Thomas Homer-Dixon warns that

> within the next fifty years, the planet's human population will probably pass 9 billion, and global economic output may quintuple. Largely as a result, scarcities of renewable sources will increase sharply. The total area of high quality agricultural land will drop, as will the extent of forests and the number of species they sustain. Coming generations will also see the widespread depletion and degradation of aquifers, rivers, and other water resources; the decline of many fisheries; and perhaps significant climate change.[53]

Many areas in Africa are already witnessing the impact of environmental instabilities such as arable land degradation, drought, and deforestation. Widespread migration, refugee flows, and conflict have set in, putting great pressure on already weak states. According to Homer-Dixon, these types of challenges, in addition to reduced fisheries, will lead to sharp increases in conflict—certainly, to more immediate increases—more so than other grave environmental changes, including ozone depletion and the long-term construct of climate change.[54]

While the developed world may not directly feel the impact of these environmental scarcities, there may be a more immediate and short-term impact felt in the near future, based on energy scarcities. As China, India, and other developing countries continue their economic growth based on the Western economic model, the competition for energy resources will grow. For the United States, the economy is a pillar of national security that depends upon vital energy resources. President Bill Clinton in 1999 stated that "prosperity at home depends upon stability in key regions with which we trade or from which we import critical commodities, such as oil and natural gas."[55] With declining reservoirs of nonrenewable resources such as oil and gas, and with past empirical evidence demonstrating that such resources are the cause of wars, the possibility of future conflict centered on these resources seems high.[56] As our cases will demonstrate, environmental insecurities crosscut all components of human security.

Case Study 1: Niger

Upon planning the trip to Niger, we were struck by the defense attaché's reassurance that Niger is a very peaceful country, unlike many of its neighboring countries. Why was Niger stable, or at least seemingly so, while the Sahel region seemed to be on the precipice of chaos? Moreover, how could a country that had just experienced a food crisis (or famine) be stable? To grasp some understanding of both Niger's commonalities with the Sahel region and its unique qualities, we will begin with a brief background, a description of the food crisis/famine, an environmental analysis of the country, an exploration of the terrorist threat, and a brief discussion of what is being done.

Background

Niger is three times the size of California, with a population of 11.3 million and growing at a rate of 3.3 percent. A continued growth rate at this level means that the population will double in twenty-one years. Niger's ethnic groups include 56 percent Hausa, 22 percent Djerma, 8.5 percent Fulani, 8 percent Tuareg, 4.3 percent Beri-Beri, and 1.2 percent Arab, Toubou, and Gourmantche. The Hausa, Djerma, and Gourmantche are primarily sedentary farmers, while most of the other groups are nomadic. Ninety-five percent of the population is Islamic. Niger is listed last in the Human Development Index (HDI), with 232 dollars per capita and a literacy rate of only 15 percent.[57]

Niger became a French colony in 1922 and was ruled by a governor general who resided in Dakar, Senegal, and a governor in Niger. Niger had some limited self-governing ability until 1958, when it became an autonomous state, and by 1960, when it became an independent state.[58] After independence, Niger was ruled by one party under President Hamani Diori. Corruption and drought led to a military coup, with Colonel Seyni Kountche ruling until his death in 1987. Colonel Ali Saibou, Kountche's chief of staff, took over and tried to liberalize Niger. He released political prisoners and established a new constitution, but the people were not satisfied. A national conference in 1991 paved the way for a transitional government as institutions were installed for the Third Republic in 1993. Although there were some successes of this republic, including a free press and free and fair elections, the economy declined. By 1996, Colonel Ibrahim Barre overthrew the Third Republic and established the Fourth Republic. However, Barre assumed the presidency under flawed electoral conditions and reversed advances in civil and political liberties. By 1999, Major Daouda Mallam Wanke overthrew Barre and established the Fifth Republic, which currently resembles the French semipresidential system. Mamadou Tandja won the presidency leading a National Movement for a Developing Society (MNSD) and the Democratic and Social Convention (CDS) coalition, and was elected again in 2004 for his second five-year term.[59]

A major challenge to peace and stability has been the conflict between the government and the Touregs. In the early 1990s, Toureg rebels championing their nomadic rights challenged the government, risking civil war. In 1995 a peace agreement ended fighting and eased the conflict. Several more agreements were signed with the help of other governments and groups, which essentially has led to greater decentralization of resource control, security management, development, reintegration of rebels, and return of refugees. The goal of the series of arrangements has been to create a climate of trust and mutual understanding among the different groups.[60] While these series of agreements have been largely successful, there are still sporadic signs of unrest in the Air Mountains, located in the north. The Azaouak Liberation Front (FLAA), comprised of some former rebels, still confronts government forces intermittently, and unfortunately, there are still land mines remaining in border areas.[61]

Most of the population lives in the south, where many relief efforts are focused. There are several potential and realized threats emanating from the south. According to one source, local fighting between herdsmen and other local communities is the primary threat.[62] The flash point is the declining availability of land and other resources, not ethnicity.[63] The dwindling availability of resources and the problems of desertification have even impacted the capital city of Niamey, and squatters, among them the most recent migrants from the impoverished rural areas, have been evicted from around the city in an attempt to protect the forests from Niger's harsh climate. One resident explained that the areas surrounding the capital have been under siege since the 1980s, and livestock breeders have had to relocate.[64] As a USAID official looked back at his first assignment in the 1970s, Niamey had 100,000 people; now it is close to one million.[65] The combination of population growth and dwindling resources makes Niamey and Niger desperate for relief.

The government has not been able to address many of these issues, though there was a brief hope that uranium would be a boost to the economy. In 1980 uranium was responsible for 75 percent of the state's revenue, but uranium's value soon plunged, causing huge deficits. These economic woes contributed to political unrest, which further reduced external aid.[66]

Interestingly, a *Boston Globe* article noted the typical images the West receives as a result of a devastated African economy—bloated bellies, dying children, and rampant diseases—but added, "To define Niger in such terms does grievous injury to the larger reality of a robust, life-affirming, and religiously tolerant people."[67] This is a key point, perhaps related to the question we first asked: why is Niger seemingly more peaceful than the rest of the Sahel region? However, one of the potential threats to Niger is the influence of northern Nigeria's political and extremist Islamic practices on the southern border region of Niger. For example, as mentioned earlier, some Nigerien women who live in the south have been told by radicals that any vaccination attempts are really attempts by the West to sterilize their children.[68] Clearly, such fallacies are detrimental to society and its subsequent political and socioeconomic development. The subjugation of women as a result of radical Islam further hinders development.[69] Thankfully, Niger, due to its history and culture, is not necessarily widely susceptible to such radical influences.

While people we spoke with noted that there are more women wearing head covers, Islam was still described as "soft." William F. S. Miles describes a key difference between Islam in Nigeria and Niger: "Unlike Nigeria, Islam in Niger does not serve as an identity marker for parts of the population who wish to gain status, prestige, power, or wealth over other regions."[70] The form of Islam emanating from Nigeria is known as Izala. It is a more decentralized, radical form of Islam. Most Nigeriens follow a Sufi form of Islam known as Tijanyya, which is more hierarchical and adaptive to the local cultural practices.[71] Why this difference in Islamic practices between Niger and the bordering state of Nigeria?

In Niger, the Sufi form of Islam coexists with several different forms of Islam, though in 1990, it was the most prevalent form under the Association Islamique du Niger, a government-sanctioned organization. Since the 1990s, there has been a proliferation of associations. The most notable association is the Jama'at izalat al-bid'a wa Iqamat as-Sunna.

This association comes from the Izala movement, "which proclaims the suppression of innovation and the restoration of the sunna, for a return to the original religion."[72] The French legacy in Niger promoted a tangible separation of the state and religion, whereas the British in Nigeria courted the religious leaders as political allies. Upon independence, Niger's leaders continued the French dissociative approach towards religion and the state and continued the state subordination of Islamic leaders and organizations. In Nigeria, Islam emerged as a political force in the north, and the Izala form particularly rejected any Islamic practices influenced by local cultures. However, in Niger, Islam continued its adaptive and "soft" nature that encouraged inclusiveness and local cultural practices.[73] Clearly, some influence has occurred along the Niger-Nigerian border. But just as Western globalization is maligned and elicits a backlash, so may Islamic globalization. Mills asks, "Will there be a Chinua Achebe, lamenting the passing of the old ways in favor of the 'enlightened' new?"[74] In short, radicalization is not definite for the future of Niger, but there are some danger signs. Probably the most significant aspect is the incessant force of globalization and how the state chooses to respond.

Is Terrorism a Threat?
It is difficult to determine the extent of international terrorism in Niger. There have been reports of the Salafist Group for Preaching and Combat (GSPC) planning attacks against Niger.[75] Incidents have been rather isolated, and it is difficult to distinguish between banditry, smuggling, and terrorist activities. What is worrisome is that many foreign fighters in Iraq have been linked to North Africa. Just as the fight against the Soviets in Afghanistan internationalized the jihad in the 1980s, there could be a similar effect once these fighters return to their homes.[76]

What exactly is the GSPC? It is a faction of the Algerian Group Islamique Armé (GIA) and formed in 1998 after the GIA's very violent period between 1996 and 1997. The GSPC, led by Hassan Hattab, continues to fight against the Algerian state, and it is the only group that has operated across all of the Sahelian countries. Although the leaders of GSPC have pledged allegiance to Al-Qaeda, it is not clear that there is much significance in this pledge. One of the ties connecting the GSPC to Al-Qaeda is Emad Abdelwahid Ahmed Alwan, who was accused of the attack on the USS *Cole* and suspected of planning a U.S. embassy attack in Bamako. He has since been killed in September 2002, but there was an affiliation alleged between him and the GSPC. There have been reported clashes between the Nigerien army and the GSPC. In March 2004, the army fought the GSPC on the border with Chad; in April 2004, the army killed traffickers associated with the GSPC near the Malian border, and in August of 2004, the army clashed with Toureg bandits.[77]

Stephen Ellis believes that the GSPC is more of a bandit-like criminal group that takes advantage of the ungoverned space of the Sahara, just as outlaws have done for years. Interestingly, they, too, take advantage of any infrastructure improvements for movement of contraband. Unfortunately, they also develop ties with officials such as border guards. Ellis further warns that such a potent mix of relationships could draw in U.S. soldiers such that they unwittingly become pawns as illicit groups vie for power and profit. Furthermore, closing down smuggling activities may not only affect terrorists but may also harm ordinary people by closing down their livelihoods.[78]

Much of the discussion about terrorism in Africa focuses on the ripe conditions for terrorists. Susan Rice, former assistant secretary for Africa affairs, said,

> Much of Africa has become a veritable incubator for the foot soldiers of terrorism. Its poor, young, disaffected, unhealthy, undereducated populations often have no stake in the government, no faith in the future, and harbor an easily exploitable discontent with the status quo...These are the swamps we must drain. And we must do so for the cold, hard reason that to do otherwise, we are going to place our national security at further and more permanent risk.[79]

But to drain the swamps, we must understand the security environment from the individual's point of view. Nothing seemed to put the desperate situation in Niger in as sharp a focus as the food crisis of 2005.

A Food Crisis or Famine?

It was interesting to note that many of the officials from the government and NGOs were adamant about referring to the food crisis as a crisis and not a famine. A famine indicates a government that cannot feed its people. According to one official, a famine reflects people and livestock dying of hunger, while a food crisis describes a condition where food reserves are empty and widespread malnutrition exists.[80] Whether the situation was a crisis or a famine, it is clear that Niger suffers from a chronic structural condition that leaves it vulnerable to periodic crises/famines, which put its population at risk for widespread health risks and death.

There appear to be myriad factors at work that put Niger in such a precarious situation. One observer notes that the problem is chronic poverty, not natural disasters, conflict, or corruption.[81] Government and UN officials maintain that the combination of drought and locust infestation caused Niger's "nutritional emergency." They also maintain, though, that the emergency was localized and resulted in an 11 percent loss of cereal.[82]

Warning signs of a coming emergency seemed visible. Médecins Sans Frontières (MSF) noted the wave of malnourished children at their feeding center in Maradi. With low food reserves and a resulting increase in cereal prices, people could not afford food or medical assistance for their families. Subsequently, people sold their livestock, which caused livestock prices to fall.[83] Another causal factor was the pace of donor response. The World Food Program (WFP) saw signs of an imminent emergency in 2004, but donor response was not quick enough. The WFP noted that government-subsidized food was not enough to combat this problem, so it established free food distribution and food-for-work programs.[84] Unfortunately, much aid did not reach the intended population. One person noted that, as we mentioned earlier, some aid drifted to Nigeria and then was resold at unaffordable prices in Niger.

The sicknesses that develop in such an emergency have second- and third-order effects that are not clearly or immediately visible. For example, when mothers bring their sick children to treatment centers, siblings may be left home alone. When large numbers of sick children converge on one location, there is risk of spreading infection.[85] According to the UNDP, 3.6 million people were affected by the food shortage, with some places experiencing a daily toll of 4.1 deaths per 10,000 people. According to Thierry Allafort-Duverger of MSF, 30,000 children had been treated for malnutrition and undernutrition midway through the crisis, and he predicted 50,000 would be treated by the end of the year.[86] Allafort-Duverger noted that "the slightest breakdown, be it a drop of harvests or a rise in prices, is enough to trigger a dramatic rise in the number of children falling victim to severe malnutrition."[87]

Moreover, the problem extends beyond Niger's borders. According to UNDP administrator Kemal Dervis, "there needs to be an increase of resources allocated to the Sahelian region. A long term commitment is needed for the region, not just in response to crisis." Furthermore, he states that "our work now has to focus on the complex mechanisms that need to be put in place to really ensure food security, reduce infant mortality, and increase maternal health."[88] Dervis points to the importance of women's empowerment and education to positively impact these conditions.[89] Another indicator of the importance of educating girls is the linkage between education and fertility rates. According to the United Nations, women with no education have twice the fertility rate as women with ten or more years of education. In another study, seven years of education seemed to correspond with fertility decline. Not only is education linked to falling fertility rates, but it provides an avenue for increased participation of women in the economy, society, and governance—all factors that lead to human development, as opposed to factors that gravely contribute to human misery.[90] Interestingly, a nutritionist from UNICEF mentioned that even teaching women the simple task of breastfeeding instead of giving babies water (usually contaminated) would go a long way in fighting health problems.[91] Kofi Annan, after visiting Niger and listening to starving villagers and seeing the effects of hunger on babies, concurred that Niger requires more than free food.[92] As a first step in understanding the complexity of Niger's condition, the next section provides an environmental analysis. With the declining environmental conditions and resources combined with an explosive population growth, human survival is precarious. If the state cannot find a way to secure people's basic needs, they will turn to whoever offers relief.

Niger: A Delicate Contrast in Climate Zones
Like many of its sub-Saharan neighbors, the country of Niger sits landlocked and precariously positioned in a region affected by the ebb and flow of shifting climate patterns. This combination makes its southern region vastly different than the north. Only the extreme southern border area of Niger with Burkina Faso, Benin, and Nigeria receives any significant precipitation, which comes in a wet-dry pattern typical of tropical savannas. While the southern 20 percent of the country falls within the definition of a tropical steppe climate, the remaining 80 percent of Niger is considered a true desert climate. Four climographs, figures 1.1–1.4, highlight the winter dry–summer "wet" pattern of precipitation typical to Niger along its southernmost border (figure 1.1) and its steady decrease in precipitation as one moves northward (figures 1.2–1.4). For this purpose, the capital, Niamey, and the cities of Tahoua, Agadez, and Bilma—the only cities with reliable climate records—are used.[93]

The area closest to Niamey (figure 1.1) shows the effects of the northward portion of the intertropical convergence zone's (ITCZ) seasonal rains. At higher latitude (figures 1.3 and 1.4), it is clear that the majority of Niger is a true desert under the influence of subtropical high pressure (STH).

The late 1960s, 1970s, and 1980s saw a catastrophic, decades-plus drought that afflicted the larger Sahel region of Africa (Niger included). While the catalyst for this drought is still debated among climatologists, one fact is certain: desertification continues to affect the lives of millions of Sahelians. Whether this desertification is anthropogenically induced, a global warming consequence, or due to a combination of factors, the implications to regional stability and security for this developing part of the world are clear, given the area's reliance on changing and increasingly unreliable climate patterns whose end state is unknown. As we have discussed, climate change is not itself the cause of Africa's instability.

Figure **1.1.** Climograph of Niamey, Niger.

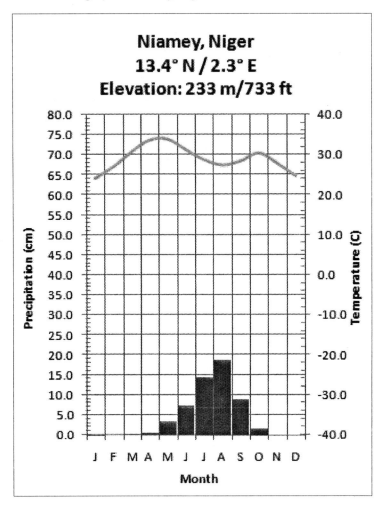

But it is a very significant component, as "uncontrollable climatic factors complicate Africa's more controllable problems, such as population
and poverty" and inevitably manifest themselves in the form of ever-
increasing tensions between neighboring states.[94]

FIGURE **1.2.** Climograph of Tahoua, Niger.

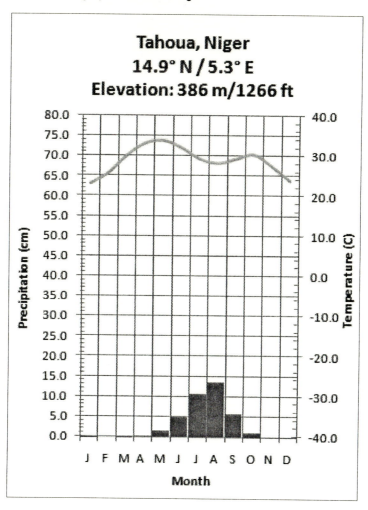

Climate Change and Desertification Within the Sahel
The notion that climate change and desertification are both underway within the Sahel region is no longer challenged, even if the genesis and responsible mechanisms are still the subject of debate. The Sahelian

FIGURE **1.3.** Climograph of Agadez, Niger.

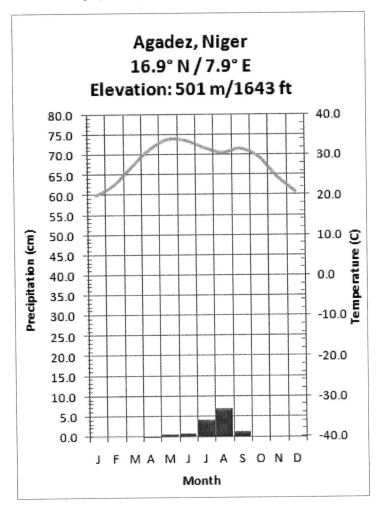

drought which lasted from 1968 though the mid-1980s is generally believed to be an example of climate variability at its most severe, responsible for many deaths and the collapse of agriculture systems throughout the region.[95] Central to the problem is the very definition of

Figure **1.4.** Climograph of Bilma, Niger.

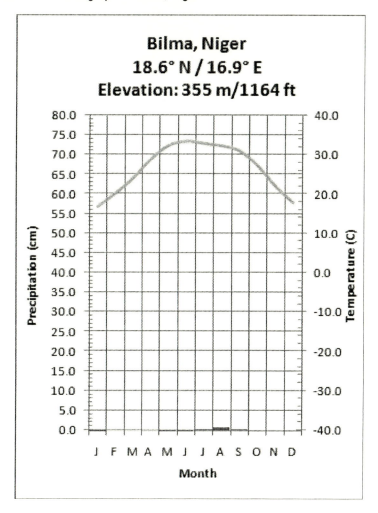

"desertification." It was not until 1992 that the United Nations Environment Programme (UNEP) defined desertification for the benefit of the
Intergovernmental Negotiating Committee on Desertification (INC-D)
as "land degradation in arid, semi-arid and dry sub-humid areas resulting

from various factors including climatic variations and human activities." Accordingly, the very definition of desertification has been refined by some to mean "land degradation in dryland regions, or the permanent decline in the potential of the land to support biological activity and, hence, human welfare."[96]

Niger's relative location within the larger Sahel region means that problems associated with accelerated climate change and desertification directly impact its ability to support its rapidly increasing and very poor population.[97] As of 2005, 38 percent of Niger's fledgling economy was dominated by rural, rain-fed, subsistence agriculture, and livestock production centered along its only arable stretch of land, the southern border.[98] The wide range of economic stressors affecting Niger, combined with its dependence on a marginal climate, makes small variations from the expected climatic norm able to significantly and deleteriously affect the lives of most of its people. The future of the Sahel is rather uncertain, especially when viewed in light of climate change considerations and the almost unimaginable set of interactions involving possible global warming, increased land degradation, and other inhibitors of human security.

The climate record over most of the African continent shows a warming of 0.7°C during the twentieth century, along with an episodic decrease in precipitation over significant portions of the Sahel and central Africa.[99] Global circulation models (GCMs) run by climatologists working for the Intergovernmental Panel on Climate Change (IPCC) have attempted to draw a link between an increasingly warmer planet (and thus, a warmer ocean) and precipitation patterns.[100] The findings show that although an increase in overall rainfall, when globally averaged, is likely, this is not the case throughout, with drier than normal conditions expected to prevail over the southwestern Sahel, specifically, Mali and Niger.[101] While computer-model reliability in forecasting long-term rainfall patterns remains uncertain, most models consistently show a decrease in rainfall over the region that accompanies a continent-wide rise in temperature ranging from 0.2°C to 0.5°C per decade through the end of the century.[102] This translates to an increase in temperature ranging between 1.8°C and 4.5°C over the next ninety years, a conclusion whose magnitude and implications are more than merely academic. Numerical model output suggests that for every one degree Celsius of mean continent-wide warming, a corresponding 6 percent reduction in rainfall will follow, bringing the potential reduction in rainfall over the Sahel region of the African continent to anywhere from 10 to 30 percent by the year 2100.[103] Understanding the potential impacts of such dire climate simulations is critical to assessing what might happen to regional stability. If continent-wide drying occurs as a slow and steady process, the ecosystem might have time to adapt and cope with the consequences. If the rate of drying, however, is more sudden and concentrated, like the 20 to 40 percent declines observed during the 1960s through the 1980s, then the whole system might be severely stressed.[104] If the latter is the case, there will likely be a large-scale failure in the region to meet the demand for subsistence food and water.

Security Implications of Climate Change and Desertification: A Niger Perspective

The long-term implications of climate change and desertification within sub-Saharan Africa will continue to manifest themselves in the form of widespread intra- and interstate conflict. As of 2002, it was estimated that over 50 percent of the population was fifteen years old or younger.[105] As limited and shifting resources come under an ever-increasing amount of stress, the volatile combination of a young and predominately poor population will make Niger vulnerable to continued deterioration of human security. Specifically, we focus on concerns about the integrity of water resources, food security, natural resources, and continued or accelerated desertification as factors impacting the overall human security of Niger.

Water Resources, Food Resources, and Their Impact on Human Security

Access to sanitary and reliable water resources is unavailable to approximately 2 billion people around the world, but is especially problematic in the Middle East and Africa. These regions considered water-scarce or

FIGURE 1.5. The Niger River Basin.

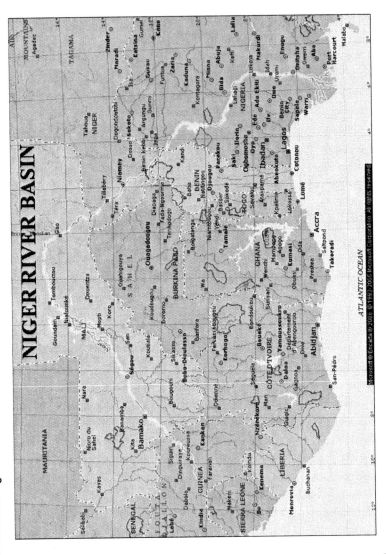

water-stressed are likely to double in area by the year 2025. Changes in climate are theorized to magnify both episodic and longer-term short-falls in water availability within already arid and semiarid locations. Since many parts of Africa, for example, are dependent on simple, non-redundant, and single-point water sources for survival, any significant dry period can be catastrophic. Within the Sahel, the ability to respond to the many potential second- and third-order effects brought about by climate change will depend on the region's ability to manage water resources from an intrastate and interstate perspective. This means placing emphasis on integrated, cross-state concerns and properly managing river basins to the greatest extent possible.[106]

Generally speaking, close to 800 million people around the world are considered undernourished or malnourished, with many of them in Africa. Climate change is likely to affect the overall agricultural production cycle by inducing changes in the growing season as a result of shifting temperature and precipitation patterns. While some changes might be beneficial, such as longer growing seasons, many other interactions may offset any gains. In regions where agricultural practices are robust and well established, the impacts of climate change might be easier to overcome. However, in places where agriculture is already at or near the climatic tolerance of staple crops, or where subsistence practices dominate, any change to the climate norm is likely to lead to complete failure.[107]

Niger is a state with only one active river basin, the Niger River. All other stream channels within its boundaries are ephemeral, making the Niger River a vital enabler to the nation's otherwise fragile economy, as well as to the entire western Sahel region. The river's headwaters originate in the highlands of Guinea near the Sierra Leone border. It runs through Mali, the border between Benin and Niger, and finally Nigeria, where it empties into the Gulf of Guinea (figure 1.5). With few interruptions, approximately five thousand kilometers of its length (including tributaries) are navigable, making the Niger the lifeblood of a sustainable fishing, trading, and agricultural economy throughout the western Sahel. However, the Niger River is a stream that, like the Sahel itself, traverses a changeable environment, from humid to arid and back to humid. Thus, it is very vulnerable to slight changes in environmental conditions.[108] Far from the river's headwaters and downstream from the point of minimal flow (central Mali), the Niger is especially susceptible to the whims of variable precipitation patterns and anthropogenic forces. Irrigation schemes dating back to the early 1930s have successively dammed and diverted water away from its main channel, while additional flow from tributaries has been impeded due to smaller waterworks projects. What water does flow into the state of Niger remains completely unregulated and largely untapped except for extremely small-scale, local-use purposes.

Hydrologic records dating back to the early 1960s show that many of Africa's major river basins have undergone a steady decrease in water discharge. The historical record for Niamey, Niger, for example, shows a decrease in Niger River flow volume by more than 10 percent. This is likely due to a combination of less precipitation, more evaporation, and higher stress on upstream water flow.[109] In places like Mali or Nigeria, where hydroelectrical plants benefit from a more reliable river discharge and reservoir level, declines to the Niger River basin flow are less problematic than in Niger, where basic irrigation needs can be obliterated over multiple seasons, affecting the livelihood of farmers and pastoralists engaged in basic subsistence practices.

As stated, Niger's relative location within the Sahel exerts an enormous amount of pressure on already weak pastoral and agricultural activities. Pastoralists from Niger, for example, tend to move their herds north or south with the seasonal rains. Often, this takes them into Nigeria across a largely unregulated and ungoverned border, where their animals can feed and survive on literally greener pastures. These crossborder activities aggravate existing tensions between the states as competition for scarce resources intensifies. Another complicating factor for pastoralists and subsistence farmers is the greater region's desire to engage in the cultivation of cash crops for the global market. This endeavor drives subsistence farmers and pastoralists into progressively marginal lands, increasing the area's susceptibility to conflict. These human and environmental competitions set up a downward spiral that further exacerbates existing human security concerns. This insecurity may prevent the institution of appropriate management practices and techniques that might otherwise prevent the loss and/or waste of valuable natural resources such as water and food. The end result would be a cycle of increased competition and continued conflict.[110]

What Is Being Done?

One of the authors was able to meet with several members of the UN, NGOs, and host nation as well as U.S. government officials in August 2006. Observations during this trip support many of the points addressed above. We had tremendous support from Lt. Col. Stephen Hughes, the defense attaché, who coordinated the meetings as indicated throughout our discussion on Niger. At this point, we will highlight some persistent themes.

First, we would be remiss if we did not mention the tremendously selfless, compassionate people we met. It is inspiring to meet people who want to make a difference and have chosen to work hard under harsh conditions, under what appear to be insurmountable odds. Herein lies the modest strategy. For the most part, there is good work happening in small doses, but over time, these efforts will make a difference. As an example, one gentleman who served as a translator regularly assisted his home village, primarily in the field of education, even though he no longer resides there. Having said this, a great deal more can be accomplished to assist this type of individual work. A far more coordinated effort among NGOs, the UN, the host nation, the U.S. government, and neighboring countries is required. This is more easily said than done, but unfortunately, each organization has a specific charter that narrows the focus of effort while creating critical blind spots that prevent anticipation of unintended consequences and negative second- and third-order effects.

Some critical themes that crosscut many of the current efforts, we find, are significant. First, as the gentleman in charge of a human rights NGO explained, there must be rule of law and justice. Second, education is critical to producing a literate citizenry with a more adaptable and resilient mind-set, which in turn can produce leaders and institutions necessary for a functioning, legitimate government. Third, education will provide tools for cultural change, specifically concerning ways of life that depend on scarce land and in terms of elevating girls' and women's participation in social, economic, and political life. The current conflicts between pastoralists and agriculturalists seen as land becomes scarcer may be abated by the diffusion of more sustainable practices which could both increase yield and decrease land degradation.[111]

There are local practices that can be leveraged to better address Niger's complex threats. For example, a cultural practice, *Habbanae*, was recognized by the Harmony List in 2006 as one that provides sustainable development and enriches quality of life. In short, the cultural practice occurs when a herdsman falls on hard times. The community responds by providing a pregnant cow for three years. This practice not only helps those in need, it also strengthens the bonds of the community.[112] There are cultural practices that also prevent adaptability. For example, people have come to accept famine as a part of life. People still grow millet, which does not provide adequate nutrition. Crop diversification, which would introduce a more nutritious, protein-enriched diet, is key to survival. Such change is difficult in a society that is uneducated and continues to experience a death rate of 262 per 1,000 children under the age of five. It is difficult to talk about family planning under these conditions, so women, on average, continue to have seven babies each.[113] But women need to be a target for programs that empower and educate. For example, a UNICEF program helps women with malnourished children by providing them with goats, especially valuable during periods of drought. Focusing on projects for women provides a "direct advantage to the children."[114]

In fact, this focus on women is one of three strategic points outlined by Mark Wentling, a career USAID officer with multiple assignments in Niger. Overall, his strategic plan calls for

- Conserving and Making Land More Productive

    –Assist farmers to use and manage natural regeneration methods in their fields
    –Transfer, as possible, to farmer fields Sahelian Eco-Farm Practice

–Work with communities to draft land use maps of their geographical areas

–Help communities obtain improved seeds, seedlings, and cuttings

–Establish, where possible, "off-season" small irrigation schemes

- Improving the Well-Being of Women and Children (Women's Groups)

–Provide nutritional, reproductive health and family planning education

–Arrange for the identification and treatment of all children suffering from malnutrition

–Support life skills (asset management) and literacy training (critical thinking)

–Organize micro-credit schemes for women

–Train women to be good managers of cereal banks

–Help women improve their livestock holdings

–Increase the number of children in school, especially girls

–Introduce tree education programs and increase the number and variety of trees women and children possess

- Improving Community Well-Being

–Help provide and maintain adequate water, sanitation and health services

–Mount and maintain a malaria prevention and control program

–Assist with improving animal health and growth

–Develop marketing plans to obtain highest prices for community/farm products

–Provide training to men in the good management of their assets and incomes

–Arrange for Food-for-Work activities during the hungry seasons[115]

Wentling provides guidance for this strategy. First, it must be viewed holistically, which requires understanding how one aspect of the strategy affects other portions of it. Second, the community must be empowered and must be a stakeholder in any project.[116] Wentling acknowledges that "some of these choices [about changing ways of life and thinking] will be very hard to make but, in the end, it is the communities and their leaders that must make them."[117]

Homer-Dixon reminds us "that resource scarcity is, in part, subjective; it is determined not just by physical limits, but also by preferences, beliefs, and norms."[118] As we explore the environment as a critical component of human security, it is important to note that there are possible cultural and technological adaptations that could provide relief in the short term and perhaps even long-term sustainability. The first step is acknowledging the short- and long-term impacts of environmental change. Human insecurities know no boundaries, so we move on to our next case, a neighbor of Niger: Chad.

CASE STUDY TWO: CHAD[119]

Chad, our secondary case, provides another excellent opportunity to understand the political and socioeconomic situation through a human security lens. Clearly, it suffers from many of the same environmental conditions outlined earlier. As a result, it, too, presents vulnerable populations to dangerous radicals, extremists, and terrorists. Our trip to Chad, however, not only allowed us to pay attention to the extremely vulnerable refugees as a result of the crisis in neighboring Sudan, but also gave us a unique opportunity to witness the tremendous work by many different organizations who are trying to alleviate the situation. Conflict prevention will be key to managing the potential crisis in eastern Chad, which is mounting as a result of the refugee crisis that looks more like a permanent situation than a temporary one. This case will focus on the refugee situation, which has only exacerbated the fragile environmental conditions with added competition for scarce resources.

Background

At first glance, Chad looks similar to Niger. It, too, is landlocked. And, though we are long past the days of ascribing to environmental determinism, geography still matters. Jeffrey Sachs notes studies by Harvard's Center for International Development that demonstrate the link between proximity to the sea and wealth. The poorest countries in the world are tropical and landlocked; those countries are Chad, Mali, Niger, Central African Republic, Rwanda, Burundi, and Bolivia.[120] Sachs attributes two challenges to international development: climate and geographical obstacles to transport. In the age of globalization, these challenges have real effects on success or failure.[121] Vast gaps in wealth between countries emerged when the world economy became industrial and science-based, starting two hundred years ago.[122] These inequalities have only grown. It is in these disadvantaged countries that we see the gravest humanitarian and social tragedies.[123]

Chad is twice the size of Texas, with a 9.5 million population. The growth rate is 3 percent—which equates to a doubling time of twenty-three years—and there exist more than two hundred ethnic groups. They include, in the north and center, the Gorane, Zaghawa, Kanembou, Oaddai, Arabs, Baguimi, Hadjerai, Fulbe, Kotoko, Hausa, Boulala, and Maba, and in the south, the Sara, Moudang, Moussei, and Massa. There are over 120 indigenous languages, but the official languages include French and Arabic (Chadian Arabic). Fifty-one percent of the population is Muslim, 35 percent is Christian, 7 percent is animist, and 7 percent is various other religions.[124]

To even begin to understand the mosaic of cultures, one must first understand the climate, political strife, and Sudanese crisis. As discussed earlier, the climate in Chad and the rest of the Sahel is one of extreme rainfall variability. Over the last forty years, overall rainfall amounts have decreased, which has had an adverse effect on the area. The region experienced a wet period in the 1950s and 1960s, followed by twenty years of low rainfall in the 1970s and 1980s, and then some increase of rainfall in the 1990s. There is not complete consensus on the reasons for these changes, which highlights the importance of further environmental research. However, it is clear that a drought in a populated area produces food crises or famines. Extreme climate fluctuations have an effect on food-collecting strategies and other economic decisions. Populations travel and may experiment with other methods for survival.[125] The profound cultural changes that accompany such changes in food systems have great impacts. According to John H. Bodley, "Food Systems are cultural mechanisms for meeting basic nutritional needs."[126] When local coping mechanisms are overrun by maladaptive and counterproductive structures of the state, then there is potential for more widespread famine and other cultural dislocation.[127] Added to these environmental changes is Chad's civil war, which began in 1965 and continued on and off into the 1990s. While the conflicts seem to have common themes of north versus south, Islam versus Christianity and animism, the food-gathering cultures of pastoralism and agriculture have played a key role as well. These civil outbreaks have come at great cost. Conflict causes people "to migrate to other areas; it destroys cultures, social networks, and leads to an insecure and uncertain economic and social situation."[128] It is no wonder that the per capita income in Chad is only 237 dollars.[129]

The political system since Chad's independence from France has been marked by strife, war, ineffectiveness, and illegitimacy. Civil war broke out in 1965 that mainly pitted the Muslim north and east against the Christian south, a coup occurred in 1975, and widespread fighting broke out again in 1979. Hissein Habre emerged as the leader until 1989, when the current president, Idriss Deby, overthrew him. In 1991 Deby became president and subsequently won the 1996 and 2001 elections. In 2004 the National Assembly approved his desire to remain in office beyond the constitutional limit of two terms.[130]

Two of the authors had just arrived in country in June 2005, when Chad held a referendum on Deby's third term. Interestingly, hardly anyone went to the polls, which perhaps indicated complete apathy or defiance of the president. Key to understanding the complexity of Chad are the president's close ties to his ethnic group, the Zaghawa, of which many members inhabit refugee camps in eastern Chad. Not surprisingly, many of his top advisors, including the military, are also from the Zaghawa clan.[131] We had a chance to speak with a leader of an opposition group, the Justice and Equality Movement, or JEM, from Sudan, who had protection in N'djamena, the capital city of Chad. According to this rebel leader, one of the platforms the JEM stands for is security for the greater Sudan. Yet, he identifies himself first as a Zaghawan and secondarily as Sudanese. It seems difficult, then, to advocate for and promote a Sudanese nationalistic agenda. This situation is reflective of weak states that have yet to elicit national loyalty and legitimacy from their citizens, whose primary allegiance is to their own ethnic group rather than to the government of a state whose motivations are suspect.

Regional Spillover Effects: Sudan

There are many estimates coming out of the Darfur region regarding the number of civilians killed and displaced. One estimate has the total at more than 2.5 million as of December 2004. Many of the reports describe unbelievable cruelty and brutality. There is local and international consensus that the perpetrators are the Sudanese military in conjunction with the Arab militias, known as the "Janjaweed" or "evil horsemen."[132] Reports indicate that the attacks began in February of 2003 by the Sudan Liberation Movement/Army (SLM/A), followed by the JEM's attacks. Then, in April 2003, the SLA launched an attack on the airport in the capital of the northern Darfur state, destroying seven planes and killing approximately one hundred soldiers. Immediately following this attack, the Janjaweed formed, armed by the government, and conducted brutal retributional attacks against the civilian population.[133] By January 2005, the UN reported that there were approximately "1.65 million internally displaced persons (IDPs) living in 81 camps and safe areas, plus another 627,000 'conflict affected persons,' and 203,000 refugees in Chad."[134]

Darfur consists of roughly 20 percent of Sudan and holds about 14 percent of Sudan's population. The population is a mixture of Muslim Arab and Muslim non-Arab ethnic groups. One of the largest non-Arab groups is the Fur ("Darfur" translates to "homeland of the Fur"); significantly, the nomadic non-Arab group is the Zaghawa (President Deby's group), and the Arab group, known as Meidab, live predominantly in the northern Darfur state. In the western Darfur state are sedentary non-Arabs from the Fur, Massalit, Daju, and other smaller groups. In a familiar theme, the climate and scarce resources have always been a source of periodic conflict, mainly between nomadic Arab herders and non-Arab farmers. Land has been a source of conflict, especially with the continued desertification over the years.[135] With a neighboring country suffering from what some call genocide, it is no wonder that border is tense. Before 2006, the Janjaweed's border activities had decreased with the increase of Chadian, French, and African Union forces' surveillance.[136]

However, there have always been reports of banditry along the border and in Chad, and in 2006, the fighting in the Sudan actually spilled over into Chad. An ongoing security concern centers on the tensions between the Sudanese refugees in Chad and the east Chadians living in the proximity of the camps. With the increased number of refugees, there is more competition for the already scarce resources in the area. We will come back to this point, but the situation will only worsen as the strain on resources, in combination with the already desperate straits of both native Chadians and refugees, continues.[137]

Terrorism

Is terrorism a real threat in such a situation? Clearly, the Sahel is being proclaimed by the U.S. military as "the new front on terrorism."[138] There is some question, however, concerning how real a threat terrorism poses. Chad, and, in general, the Sahel, suffers from unhealthy societal-political relations. Governments, especially in Chad, are corrupt and ineffective and, most importantly, are illegitimate in the eyes of the people. There should be concern of other dangerous actors exploiting this vulnerability. It is important that the United States and other members of the international community analyze terrorism in the context of the larger security situation facing the local people.

According to the International Crisis Group, "the Sahel is not a hotbed of terrorist activity. A misconceived and heavy handed approach could tip the scale in the wrong way."[139] For more than sixty years, the Sahel has experienced fundamentalist Islam that did not have ties to anti-Western violence. In fact, the much-cited Salafi group, Preaching and Combat (GSPC), which has its origins in Algeria, suffered a grave defeat by Algerian and Sahelian forces. It crossed into Chad in 2004, chased by the Pan-Sahelian Initiative's trained forces, and battled the Chadian army. In the battle, the GSPC lost forty-three of its members. Thus, it would seem that the proclivity of the Sahel toward terrorist activity has been overstated.

Nonetheless, it appears that in the absence of other U.S. government agencies, the U.S. military has the resources to engage in the region and focus on the potential for terrorist activity, and that it would be prudent to do so. It is harmful if U.S. military planners understand the region's problems only by viewing the complex issues of the Sahel through a counterterrorism lens.[140] This overly simple view is not helpful and fails to consider the many factors of human and environmental security at work here.

The ICG report warns that the governments of the Sahel may be manipulating U.S. fears of terrorism to leverage benefits, reminiscent of the Cold War days.[141] The United States must be careful not to overgeneralize about Islamic militants or extremists. For example, in the Sahel, the clerics, scholars, and political leaders focus on the inclusivity of Sahelian Islam, "pointing to the role that Sufi Brotherhoods have played in forming the regional culture of tolerance."[142] This culture of tolerance should be valued and could assist with improving human security conditions.

What's Being Done
We were in Chad in June 2005, and our observations support many of the points addressed above. We had tremendous support from the defense attaché, Lt. Col. Tim Mitchell, who enabled meetings with several members of NGOs, UNHCR, the host nation, the U.S. State Department, and the African Union and with the rebel leader from the Justice and Equality Movement. Additionally, we had follow-up conversations in Washington, DC, with members of U.S. Agency for International Development and the State Department's Conflict, Reconstruction and Stabilization Office. We had the opportunity to travel outside the capital

of N'djamena to the town of Abeche and refugee camps in Farchana and Gaga. One cannot address Chad without addressing the Sudanese crisis. As already mentioned, our observations confirm our research. The refugee crisis has exacerbated the conditions for the Chadians, and, interestingly, the UNHCR, NGO, U.S. Department of State, military, and other efforts have both helped and hurt the Chadians. The main problem is that with a corrupt Chadian government and an illegitimate Sudanese government, the organizations that are trying to alleviate the plight of both refugees and eastern Chadians cannot engage in sustainability or capacity-building activities. Instead, they are forced to provide aid directly to the people, which is beneficial but is not providing a long-term, systemic solution. In other words, the crisis seems to be the status quo. Refugees do not want to go home since it is not yet safe to do so, and the Chadian government cannot be trusted to use support that will benefit society.

What is remarkable is that we found tremendously dedicated UNHCR and NGO professionals working under extremely harsh conditions, knowing that their efforts were not solving the crisis, but yet, they were not deterred from providing direct aid to refugees who found themselves in desperate situations. The UNHCR has a charter to address refugees, while other organizations are designed to address the local population. Unfortunately, human insecurities do not discriminate between refugees, internally displaced persons (IDPs), and other categories of people. The problem that the UNHCR inadvertently created was the resentment felt by the local population suffering from human insecurities but who did not qualify under the UNHCR charter to receive aid. Fortunately, the UNHCR recognized this problem and was able to earmark some funds for the local population while seeking help from other organizations.[143] The Chadians who live in desolate regions of the country suffer from some of the same insecurities as the refugees, without the assistance of their government. They are left to cope on their own, as the extent of the Chadian government's capability lies within the city limits of its capital city of N'djamena. These ungoverned spaces are ripe with human insecurities described earlier, and, unfortunately, the people suffer.

Sustaining refugee camps is an enormous challenge. Within the refugee camps, there are security, social, education, and myriad other issues that must be dealt with on a daily basis. With everyone with whom we talked, physical security was a priority. The African Union's military forces from various African nations are charged with providing the external physical security of the refugee camps. The Chadian military units provide the internal physical security. One of the biggest challenges these military units face is the distrust or reluctance to trust by the refugees. From the refugees' perspective, how can they trust someone in a military uniform when the very people from whom they fled were wearing military uniforms? Slowly, the African Union is learning how to deal with such perceptions and taking steps to build trust.[144] Though their strengths and capabilities are very limited, this is an important first step. The UNHCR, as the primary coordinator of all the formal and informal efforts, does everything it can to work with the African Union. Cooperation between the two organizations is mutually beneficial and in the best interest of the refugees. Refugees need to feel protected, as do the UNHCR, NGO, and other personnel.

A problem arises as refugees become reliant on this secure environment and their desire to return home disappears. Most of the refugees in the camps are children (64 percent) and women. This reluctance to leave the camps becomes even more evident when the refugees are provided the support to confront poverty, famine, malnutrition, diseases, and illiteracy. Many refugees will decide to stay in the camps. In fact, if more permanent structures are erected, then the expectation for a prolonged stay increases. The reliance on receiving rations and shelter increases with each passing day. Parents willingly accept free vaccinations and education for their children. To protect the women from the perils of the surrounding Chadian villages, they are trucked to areas where they collect firewood and other resources for cooking. Other services coordinated by UNHCR include food, health, veterinary services, sanitation, water, shelter, and other critical services. With each service, cultural considerations must be understood and integrated. Some examples include the number of people culturally acceptable to live in a tent, use of centralized latrines, and modern medicine versus traditional healers.[145] For many refugees, there is no home to which to return. Their whole social fabric is gone, and the new social network is the camp. Wisely, the camps try to preserve some tribal unity in how the living quarters are arranged. With each passing day, the refugees reach a sense of routine normalcy, where they feel secure from the physical and nonphysical threats to security. The problem is that the mental attitude required for sustainment and self-help is not nurtured under these conditions. While the UNHCR and NGOs are effective in providing crisis assistance, they are aware of the long-term maladaptations that emerge.[146]

Indeed, cooperation between UNHCR, NGOs, the African Union, the U.S. Department of State, the host nation, and the military are necessary and vital to the human security of the refugees. All of the entities recognize the importance of addressing human insecurities and the protection and preservation of life as critical components of regional stability (with the exception of some of the host nation elites, as described earlier). Each organization contributes a specific set of skills and knowledge that are critical for accomplishing the common goal. What helps in the synchronization of effort is UNHCR's role as the primary coordinator. It works with the Chadian government as well to resolve issues at the national and local levels. The director of the UNHCR effort in Chad routinely involves all available NGOs in the decision-making process whenever possible. She fully understands the diversity of skills and resources NGOs can contribute. Together, they form networks of networks able to provide the flexibility and responses needed in a complex and challenging refugee and humanitarian assistance operation.

The short-term success of the efforts in the camps and outlying areas is easy to measure. One can easily determine numbers of people receiving medical aid, schooling, food, and so on. However, what are not easily measurable are the long-term impacts, especially when host nation support is minimal or even working at cross-purposes. A gentleman from the United Nations Development Programme stressed the need for elite accountability. Without changing the elite mind-set, there is little hope of permanent change.[147] Another key indicator for long-term change is education for girls. Education, especially for girls, as with the previous case, has potential for long-term cultural changes that could generate positive social, economic, and political development.[148] What was evident through our discussions is that there is an emerging consensus that human insecurities must be addressed with a holistic approach and that real, focused effort is attainable by all agencies and organizations, both military and nonmilitary, if there is predeployment planning so that everyone understands each other's roles and missions.

Our focus was on eastern Chad due to the effects of the Sudanese crisis; however, Chad experiences human insecurities throughout the country. When we discussed the northern border with one official, he recounted how U.S. forces helped the Chadian military conduct border patrols along its northern border with Libya. Unfortunately, the border patrols also stopped the smuggling upon which many ordinary Chadians rely for their livelihood. In the south, Chad also was affected by refugees of the crisis in the Central African Republic. How do these situations and conditions relate to terrorism? In truth, it is not clear; however, they present tremendous vulnerabilities within society in the midst of an illegitimate regime. As we discussed earlier, when people are faced with desperate situations that the state cannot or will not address, they may turn to alternate sources of relief. Currently, the international community is working to provide some relief. Unfortunately, without a legitimate and effective government, this relief lacks sustainment at the national level. In a harsh environment, with dwindling resources and bad governance, the added pressures of refugees only makes the competition for scarce resources more intense and conflict-prone. This situation ought to cause the world concern on many levels.

ENVIRONMENTAL POLICY IMPLICATIONS: THE CONTINUED OR ACCELERATED DESERTIFICATION AND ITS IMPACT ON HUMAN SECURITY

For both Chad and Niger, the continued degradation of the environment will only exacerbate current human insecurities. For Chad, the refugee situation is, at best, a semipermanent situation that competes for already very scarce resources, and for Niger, the food supply is intricately related to rainfall variability, among other factors. Of course, the earth's climate has been under constant change. For example, the area of the southwest United States known as the four corners region used to be a much more humid, savanna-like climate approximately 11,000 years ago. The native Pueblo peoples who once thrived in this region by cultivating a set of staple crops suddenly vanished due to a sudden and unrecoverable change in the expected environment. The region was dramatically transformed and became the large desert that is today's Colorado plateau.[149] The Sahel faces a potentially similar fate as deserts continue to expand and climate change inexorably changes temperature and precipitation in ways that will be less sustainable for life.

Continued or accelerated desertification has both natural and human components, each of which can diminish or reinforce the other. Basic temperature and precipitation parameters drive not only vegetative cover but, in turn, soil formation, types of agriculture possible, and a region's overall carrying capacity. Climatic factors can induce positive feedback cycles resulting from, for example, the loss of vegetative cover and a subsequent loss of moisture and increase in solar energy reflection (albedo), leading to progressively drier conditions. For example, loss of vegetative cover can lead to loss of soil through erosion and a decrease in precipitation, which can then lead to further loss in soil.[150]

Precipitation patterns in the Sahel can be affected by factors geographically far removed. For instance, sea surface temperature and atmospheric pressure anomalies in the Pacific and Atlantic oceans have been shown to alter the amount of rainfall over the African continent.[151] These teleconnections clearly show that the problem is indeed a global one, and that changes to environmental conditions far from the Sahel have tremendous impact on whether the climate there is favorable or hostile to humans.[152]

Anthropogenic forces driving desertification include overreliance on unsustainable agricultural practices, overgrazing, and deforestation—all of which can radically alter the Sahel's delicate ecosystem through

excessive erosion as well as loss of moisture via destruction of protec-
tive tree canopy and productive arable land. The United Nations has
noted that, on average, two-thirds of the already desertified land in Africa
became that way due to overgrazing, while the remaining one-third was
due to improper agricultural practices and deforestation.[153]

Whether anthropogenically or naturally induced, desertification reduces
soil fertility, livestock and agricultural yields, and water-retention capacity. If the Niger portion of the Sahel is rendered useless, the net result will
be an increase in interstate as well as rural to urban migration, placing
more pressure on local, city, and state-level governments already stressed
by the steady arrival of the dispossessed to places without the economic
bases and infrastructure to support them.[154]

METHODOLOGIES FOR IMPROVEMENT AND CHANGE: A MULTISTATE APPROACH

Many factors place the large majority of the African continent at an ele-
vated risk to the impacts of climate change and desertification. Given the
Sahel region's weak economic and political standing, it might be tempt-
ing to argue that integrated solutions are simply beyond its means.[155]

As a result of the known vulnerability of the region to small changes
in environmental conditions, one possible solution is the use of existing
climate models to enable the African Union in assisting states within
Africa. These models, although far from perfect, are potentially valuable
to agents within Sahelian governments. Potential uses might include
the public broadcast of climate outlooks that could enable pastoralists
and agriculturalists to better prepare for upcoming seasons. Some of
the teleconnections mentioned earlier can be forecast with some degree
of accuracy, thereby pointing to the right call to action needed. These
products are often used within the United States and can help prepare
populations if implemented in a timely manner. Specifically, climate
models run by the National Oceanographic and Atmospheric Administration (NOAA) assist U.S. farmers by predicting where a drier/wetter
than normal pattern is likely. In the western Sahel, where monsoonal

weather patterns are easier to predict, using such tools might be quite beneficial and adaptable to residents' needs. Many climate models can, and often do, take into account large-scale teleconnections, including vegetative cover, sea surface temperature, soil moisture, and other variables, in a comprehensive attempt to obtain the most accurate long-term forecast.[156] Although the infrastructure needed to undertake, deploy, and manage such a lofty technical and computational goal within the Sahel is intensive in terms of staffing and capital, such tools already exist within government agencies and educational institutions of the United States and European Union. These items could be included as part of a "climate aid package," potentially kick-starting both interstate cooperation and a larger scientific presence by climatologists, soil scientists, and others willing to lend their expertise.

Managing the already fragile water resources is a potentially contentious proposition, but it is critical. Niger, which has no ambitious water projects, must coordinate with Mali, its upstream neighbor, to establish a compact that would allow for a more reliable water flow into Niger. Without greater flow, no amount of aid or technology will allow for even the most modest irrigation projects to be emplaced. Of course, anything that happens in Niger will affect its downstream neighbor, Nigeria. At a regional level, states within the Sahel ought to consider a project similar to one proposed in southern Africa, which suggests diverting Zambezi River water from a surplus region to a deficit region.[157] Of course, such a compact, whether undertaken in the Sahel or in southern Africa, would require careful interstate cooperation, and thus is fraught with potential problems—but also possibilities.

While an agreement to divert and more formally manage the Niger's flow may seem too ambitious, the larger Sahel region, including Chad, is home to several other river basins, such as the Senegal, Black Volta, White Volta, Benue, Chari, and Hadejia. Each of these basins, in some way, traverses changeable climate regimes. The idea of closely monitoring and regulating water flow as a commodity to be sold and bought across the Sahel would seem to be a reasonable and cost-effective way of diminishing future water tensions, especially if climate change forecasts

are correct and the area is about to undergo a significant amount of drying. This project, if realized, might have a similar effect to that of the
many Colorado River projects since the early 1900s—the growth of
year-round agriculture in regions of California and Arizona that otherwise receive very little precipitation.

Once the larger-scale issues of water conservation and management
are addressed, then local and state governments can work to institute
other, less invasive techniques that promote conservation and careful
management of water resources. Many of the tribal regions within the
Sahel have long-established patterns and techniques of dealing with the
seasons. It seems appropriate and prudent that an approach combining
the state of the art in scientific thought, sound governmental policies,
and local ingenuity congruent with indigenous knowledge might be the
best integration tool for dealing with the ongoing changes to the environment at all scales of action. The long record of computer model runs and
historical data strongly suggest that climate change increases not only
uncertainty but also the chance of greater short-term variability, imply-
ing that wet regions might experience too much rain while dry regions
might experience more severe drought. If one lives in the Sahel, this
level of uncertainty can delineate the difference between life and death.

Policy Implications[158]

The patterns and insights revealed through a human security paradigm
offer considerations for policy makers. A key discovery is the recognition
of the interrelationships and connections among these various human
insecurities and actors in Chad and Niger. Addressing the issues of basic
survival at the individual level highlights the security implications at the
state, regional, and global levels. Most notable are the acute implications for weak or developing states and the regional neighborhoods in
which they reside. As states are unable or unwilling to provide security
at the individual level in the ways just described, people lose faith in
their governments and look for other ways to ensure survivability. While
sometimes people turn towards ethnic groups, kin, and other substate

entities, they may also be vulnerable to terrorist groups who offer hope and tangible support. Security policy requires a holistic approach that views human security as a critical linchpin to building state capacity and legitimacy vis-à-vis these other groups. Our cases further demonstrate how the environment directly impacts human survival. Conditions of dwindling resources and growing populations can no longer be ignored or deferred to another day.

Legitimacy, as a form of interdependence and interconnection between political systems and societies, is central to fostering good governance. Good governance provides the foundation for building state capacity to address these human insecurities. Capacity building must be inherent in any policy that will have long-term effects. In Chad, the government is so corrupt that organizations work directly with the people to ensure that they receive support. The good news is that people get aid, but the bad news is that there is no capacity-building effort in effect. Consequently, it appears that Chad, especially concerning the Sudanese refugees in Chad and the local population surrounding the camps, will be in a permanent crisis mode. For Niger, the food shortages cannot only be addressed when the cameras are rolling; there must be a renewed effort towards structural change that will foster real sustainability.

For change to occur, not only must there be significant focus at the grassroots level, but there must also be focus at the elite level. In other words, at the political level, the international community cannot turn a blind eye towards rampant state corruption and self-declared dictators. Diplomats must be empowered to courageously speak truth to power, and state officials and politicians must be held accountable. Measuring effectiveness towards the fight against terrorism will require a more nuanced standard. The United States must be careful not to fall victim to the Cold War manipulations of the past. Respect for human rights, efforts towards addressing human insecurities, and other efforts towards real political reform must be considered as standards for fighting terrorism. If not, aid, especially military aid, may mistakenly be provided, which may only further exacerbate the problem. In effect, such aid only helps militarize a regime, not democratize it.[159]

The issue, then, is how to break the mind-set that condones extreme state corruption, perpetuates human insecurities, and continues to present vulnerabilities that terrorists can exploit? How does one craft policy that empowers people and governments to take responsibility and move forward, especially in extremely harsh environments and climates? The key will be education, and it is important to realize that although the effects of education will be long term, it will be at least a generation before those effects are realized. Key to the education process will be the inclusion of women for many reasons, not least of which is that it doubles the talent pool of a society. It is important for the international community to empower local people and governments and cultivate stakeholders on all projects by working collaboratively during the decision-making and implementation process with local and national politicians and informal leaders.

Our discussion of human insecurities highlights the importance of understanding the environment. Unfortunately, the regions that are gravely suffering from human insecurities exist in harsh and extreme climates. The prognosis for Niger and Chad is that desertification will continue, thus, the urgency for finding ways to mitigate its effects is paramount. Many people living in these harsh climates face the possibility of losing their culture and their way of life. In Niger, land scarcity endangers both pastoralists and agriculturalists. This is important on several counts. First and foremost, as natural environments change, they contribute to the dynamic human insecurity challenges. These challenges put extreme pressure on governments to respond. A government's failure to respond may precipitate other adverse effects. For example, increasing scarcity of resources may lead to conflict while creating further societal, economic, and political pressures. Environmental studies that provide policy makers with forecasts and recommendations will become increasingly important as people adapt. Moreover, infrastructure becomes a key component in any policy that seeks to develop an economy. It is hard enough growing crops, but just delivering the goods to market in a timely and safe fashion can become an insurmountable task. Understanding the environment is a first step towards understanding the totality of challenges the population faces as well as the state and external actors.

Understanding these challenges will better enable us to anticipate second- and third-order effects of external actors' actions within a state and their spillover effects on the larger region. There will be unintended consequences, but by understanding these complexities, we may be able to better adapt. Our cases highlighted some of these consequences: in Chad, setting up refugee camps increased tension between eastern Chadians and the refugees as they competed for scarce resources; after the United States trained the Chadian army on effective border control techniques along the Libyan-Chadian border, unanticipated tension arose, as many Chadians had been relying on smuggling activities for their livelihood that were now shut down; and food aid targeted for Niger ended up in Nigeria, which in turn sold the aid back to Nigeriens at unaffordable prices. Understanding the integrated and complex nature of all aspects of an environment requires in-depth knowledge at all levels to mitigate these and other unintended consequences.

This enormous task requires culturally knowledgeable and aware security professionals. For the U.S. military that is currently transforming, it is clear that the most important asset that requires transforming is the intellect. The military cannot transform effectively without ensuring proper education and training of its service members and true "buy in" to the components of human security by its leaders. The foreseeable future will require adaptable, innovative, and culturally aware military members. The U.S. military has excellent professionals, but it must continue to invest at a more rapid rate in their intellectual capital. In addition, training and education must have interagency, NGO, and IGO involvement. While it is important to gain cultural perspectives of a region, it is just as important to understand the different organizational cultures among the suite of actors involved in addressing human insecurities. Furthermore, establishing an interagency, international, and military planning cell prior to arriving in a country would be invaluable. At the very least, people would better understand the different players' roles and capabilities, and many misunderstandings could be resolved.

What we discovered in Chad is that there was consensus among members of various NGOs and IGOs and host nation officials that, in fact, the military does play an important role in addressing human insecurities. The most important role is providing security where none exists. However, the military can also play a critical preventative conflict role. The Trans-Sahel Counterterrorism Partnership (TSCTP) is designed with this preventative role in mind. Interestingly, the Combined Joint Task Force (CJTF) in the Horn of Africa may be a good model, as it demonstrates how fighting terrorism with schools, bathrooms, and roads can prove effective in creating and sustaining a relationship and connection with the people in the regions of the Horn of Africa.[160] The key is basing all actions on the premise that the United States is not in the Sahel to impose an American culture. Instead, the key to success is providing empowerment to help the people to self-actualize, self-determine, and self-govern. Second, the best type of military training is professionalizing host nation militaries. In other words, educating militaries on standards of behavior will go a long way in areas where people are socialized to fear the military uniform—and for good reason. Professionalization in the long term will be more important than teaching perishable skills at the tactical level.

Fortunately, we are starting to see the interagency play a more prominent role. In Niger, USAID opened its first office in several years, though while we were in Chad, the USAID had no presence. However, the interagency elements must be resourced and empowered to bring to bear U.S. diplomatic, economic, and informational elements of power. It may be more helpful for all agencies to be organized based on regions, not states, as many of the issues we described easily spill over state borders. Perhaps it is time to create regional-level ambassadorial positions. Second, we were overwhelmingly impressed by the dedicated, selfless UNHCR and NGO professionals on the ground. However, they, too, require more resources, and unfortunately, they have been put in a crisis situation that appears never-ending unless the international community puts pressure on dysfunctional regimes. The UNHCR and the NGOs that it coordinates have been unable to create sustainable capacities in Chad because of the corrupt regime. Third, all agencies, including NGOs, must be held accountable. William Easterly criticizes the Millennium Development Goals (MDGs) because they require collective responsibility, which, in truth, means that there is no one to hold accountable. Instead, he calls for feedback and accountability on discrete tasks conducted by each organization.[161] Mechanisms for accountability are essential.

Policy makers should not use only a counterterrorism lens to view the security environment. Instead, understanding the security challenges through a human security lens provides policy makers with insights that are vital for effective policy. Attacking extremists and radicals by addressing human insecurities can be effective, although it will take time and results will not be immediately visible. Cultivating stable states that demonstrate good governance and legitimacy takes time. Many of these states are still in the early steps of state making and are struggling with enormous pressures, as discussed throughout this paper. World leaders need a better way to measure success and incorporate strategic patience in their planning and assessments. The Western concept of time will not do. For the United States and allies, this new orientation requires political and societal support as well as politically courageous and enlightened political leaders.

Policy makers must also realize that they cannot make choices for people or governments. The communities in Niger must decide their path, and for some, that may mean a shift in their thinking and way of life, especially as they develop ways to adapt to the environment. Indeed, policy makers can help create conditions that will assist people in making sound choices. These efforts will take time, intelligence, critical analysis, patience, and empathy. Subsequently, the modest strategy for success will depend on the incremental, tireless, and focused work of the international community to best succeed in creating a more peaceful and prosperous world.

CHAPTER 2

ENVIRONMENT AND HUMAN SECURITY IN THE HORN OF AFRICA[162]

Stephen F. Burgess[163]

INTRODUCTION

Human security in the Horn of Africa is as tenuous as in any region of the world, as evidenced by periodic famines, extreme poverty, and struggles over resources by growing populations.[164] Any assessment of human security in an unstable and famine-prone region like the Horn is predicated on a holistic approach that addresses causative factors such as environmental degradation, rapid population growth, and conflict.[165] Environmental degradation has caused human insecurity in the Horn. The issue is the extent to which human insecurity in the Horn is caused by other factors, particularly conflict. A secondary issue is the extent to which environmental degradation causes conflict. Is environment the primary cause of human security?

Therefore, enhancing human security posits a set of options intended to prevent struggles over scarce resources and bring sustainable development.[166] This approach is especially salient in the Horn, because the region combines high levels of environmental stress and human insecurity as well as civil wars, communal clashes, and other forms of conflict.[167]

The counterterrorism strategy of the United States in the Horn is based partly on the assumption that environmental degradation helps worsen human security and that impoverished and displaced people are more likely to support Islamic extremism and terrorist movements and create ungoverned territories.[168] Support for Islamic extremism and terrorism was manifested in the planning and execution of the 1998 U.S. embassy bombings in Kenya and Tanzania. The link between environment and human security has helped to inform the operations of U.S. government agencies in the Horn, especially the U.S. Agency for International Development (USAID)[169] and the civil affairs "hearts and minds" operations of the Combined Joint Task Force-Horn of Africa (CJTF-HOA).[170] CJTF-HOA operations, including the drilling of wells for pastoralists, have drawn criticism from development experts as "naïve" and unsustainable. In response, the United States has been developing a "3D" (development, diplomacy, and defense) approach in the Horn that aims to be holistic, focus on sustainability and human security, and include host government and NGO partners alongside U.S. government agencies.[171]

The purpose of this chapter is to analyze problems of environmental sustainability and the impact that they have on human security in the Horn of Africa. It assesses the extent to which environmental degradation causes instability and conflict. Through qualitative analysis from field research, the chapter provides an understanding of the relationship between environment and human security. In so doing, it demonstrates the impact that environment makes on human security and the effect that both have on popular support for Islamic extremism and terrorism.

A second purpose of the chapter is to evaluate U.S. efforts, especially in Somali pastoral areas in Kenya and Ethiopia, to mitigate environmental degradation, conflict, and human insecurity as well as support for extremism. In particular, it assesses the United States' "3D" approach and the CJTF-HOA model to "win hearts and minds" by assisting pastoralists in pursuing sustainable development and human security. Selected policies, programs, and actors for overcoming environmental degradation and human insecurity are examined and assessed.

HUMAN SECURITY AND ENVIRONMENT[172]

"Human security" refers to individuals, families, and communities having sufficient shelter, food, health, and other basic needs. It also means freedom from environmental degradation, famine, and disease as well as from human rights abuses, exploitation, and conflict. The concept of human security implies that "state security" does not necessarily guarantee security for individuals (especially women), families, and communities (including minority groups). The concept gained currency during the 1980s and 1990s as internal conflicts that affected large civilian populations were rising exponentially and contributing to famine, deadly diseases, and the like. In 1994 the United Nations Development Programme published the first annual volume of the *Human Development Report*, which measured human security in most UN member states.[173] Some of the greatest examples of human insecurity were the famines in Ethiopia in 1985 and Somalia in 1992, which were caused by a combination of environmental degradation, drought, conflict, and failing states.[174]

Environmental degradation refers to soil erosion, water shortages, and desertification[175] as well as to the negative effects of rapid population growth and climate change.[176] For one, climate change has quickly become a major environmental problem in Africa, where it has been contributing to severe drought and desertification as well as to widespread flooding.[177] Global warming has contributed to marked desertification in Darfur; in turn, desertification heightened the struggle between Janjaweed pastoralists and Darfur farmers over land. The struggle ultimately led to ethnic cleansing, genocide, and millions of refugees and internally displaced persons (and massive human insecurity).[178]

Population growth has long been considered one of the most important factors contributing to environmental degradation. In the Horn and other parts of Africa, population growth rates of above 3 percent per annum for the past five decades have caused mounting resource constraints. Increased amounts of people and animals have resulted in greater amounts of stress on land and water and have precipitated desertification and water shortage. Population growth contributes to communal competition and conflict over resources, which produces displaced persons, usually in the thousands or tens of thousands. In turn, the movement of hundreds of thousands of displaced people and refugees to camps has contributed to environmental degradation and periodic famine in Darfur, Somalia, and other parts of the Horn.[179]

Desertification (i.e., degradation of drylands) has been the end result of poor cultivation and pastoral practices as well as population growth and expansion and the overstocking of livestock herds. Overstocking and poor farming practices have led to soil erosion. The lack of rainfall over a sustained period of time has further contributed to desertification, which further harms human security. Global warming has added to the stress on land and rainfall that leads to desertification (and human insecurity).[180]

Water shortage as an environmental issue ranges from the macro level, which includes struggles over large rivers and lakes,[181] to the micro level of wells, boreholes, and small-scale irrigation.[182] The availability of freshwater is one of the main challenges facing the world and the Horn of Africa, in particular, in the twenty-first century. Growing populations of people and animals are depending on a finite or even shrinking amount of freshwater. Population growth, which is on track to create a world with 9 billion people in the coming decades, coupled with the negative effects of prolonged droughts and heat waves resulting from global warming, will exert increased pressure to better manage the available water supply to meet the needs of the world's population, especially the people of the Horn of Africa. Competing forces trying to maximize their use of available water may result in violent conflict if not managed properly. Water shortages are widespread in the Horn and other areas of Africa and throughout the developing world and are a major source of human insecurity. Lowered water tables and exhausted wells have disrupted both farming and pastoral activities.[183] Competition over river water is increasing, threatening the degradation of large-scale and small-scale irrigation systems and creating problems in the creation of new irrigation systems in Egypt, Sudan, Ethiopia, and other countries.[184]

Famine has killed millions in the Horn of Africa;[185] it has been the end result of severe drought, poor agricultural practices, rapid population growth, conflict, and the inability of traditional farmers and pastoralists to adapt. The inability of weak states and underdeveloped economies to prevent or mitigate famine is another factor. Conflict has disrupted arable farming and pastoral activities and has contributed to famine, especially in the Horn. The susceptibility of populations to diseases, such as malaria, cholera, measles, diarrhea, and HIV/AIDS, and poor public health services worsen the effects of famine on populations (and devastate human security).[186]

Besides the impact of environment on human security, the other major factor to consider is conflict. The literature on the interrelationship between environmental degradation and conflict and, by implication, human security takes three different positions. The first has been termed "neo-Malthusian,"[187] which draws a direct correlation between environmental stress and conflict. Michael Renner has argued that environmental stress and the conflict over resources is the greatest security threat since the end of the Cold War. The neo-Malthusians point to Darfur as prima facie evidence of the relationship between environmental degradation and conflict and deteriorating human security.

Skeptics have urged caution against the "neo-Malthusian" tendency to draw direct causal linkages among environmental scarcity and conflict and human security. Nils Petter Gleditsch, a leading scholar in peace research, has demonstrated empirically that the links are not as strong as neo-Malthusians claim.[188]

Thomas Homer-Dixon, in his 1999 book, *Environment, Scarcity, and Violence*,[189] specifies the relationship among environmental degradation, the struggle over resources, and conflict and human insecurity. He cautions against assuming a direct relationship between degradation and conflict. Homer-Dixon identifies three sources of resource depletion and degradation: supply-induced, demand-induced, and structural scarcity. Supply-induced scarcity results from a total decrease in the amount of a given resource available for consumption and is related to technologies (pumps) and practices (irrigation) used in the consumption of the resource. Demand-induced scarcity results from an increase in total population and other changes in consumption patterns. The third type, structural scarcity, is caused by a "severe imbalance in the distribution of wealth and power that results in some groups in a society getting disproportionately large slices of the resource pie, whereas others get slices that are too small to sustain their livelihoods."[190]

Structural scarcity has been a factor in nearly every case where resource scarcity has resulted in conflict. Generally speaking, none of these factors operate alone; all interact and reinforce each other in negative ways. For example, if a rapidly growing population is dependent on a fixed amount of cropland, the demand pressures (requiring more food production per acre) will result in overfarming of the available land, which reduces the fertility of the soil, which becomes incapable of producing the required yield to support the population (a supply-induced scarcity). Any outside pressure from an elite group (or warlords) that limits the available land to farm, such as government regulations or a feudal system (structural scarcity), exacerbates the shortages. Homer-Dixon warns that "environmental scarcity is never a sole or sufficient cause of large migrations, poverty, or violence; it always joins with other economic, political, and social factors to produce its effects."[191]

Disputes resulting from the scarcity of river water (such as the scarcity of Nile River water) cause "resource wars," according to Homer-Dixon,[192] who lists four conditions that must be met in order to produce conflict over limited water resources, specifically rivers:

> The downstream country must be highly dependent on the
> water for its national well being; the upstream country must be

threatening to restrict substantially the river's flow; there must be a history of antagonism between the two countries; and, most importantly, the downstream country must believe it is militarily stronger than the upstream country. Downstream countries often fear that their upstream neighbors will use water as a means of leverage. This situation is particularly dangerous if the downstream country also believes it has the military power to rectify the situation.[193]

River water produces a potential conflict between upstream and downstream states. The downstream state is at the mercy of upstream states and must be prepared to coerce upstream states if supply is threatened. The imperative increases as states modernize and use more of the resource. However, the tendency of downstream states to develop water systems first, claim "senior rights," and seek hegemony over water resources causes difficulties in attempts to establish cooperation.[194]

In regard to conflict over rivers, the trend indicates that the pressures leading to conflict are greater today than in the past. Of the twenty-four disputes over rivers, twenty-three have occurred since 1947, and thirteen (or more than half) have occurred since 1974.[195] Therefore, the frequency of interstate conflicts involving water has escalated and points towards future conflict.

A Peace Research Institute of Oslo (PRIO) study has provided confirmation of the links among environmental stress, competition over resources, and nonstate conflict in Africa. Resource scarcity has been found to cause smaller-scale internal conflicts with no direct state involvement, taking into account rainfall, population growth, and level of development. Marginalized regions run a higher risk of nonstate conflicts, and the level of conflict is affected by decreasing access to renewable resources.[196]

The chapter examines the hypothesis that increasing levels of environmental degradation are related with greater human insecurity. The hypothesis that environmental degradation causes conflict, which in turn causes human insecurity, is also examined. The relative importance of demand, supply, and structurally related shortages is assessed.

The comparative significance of macro- (or structural), intermediate- (or institutional), and micro-level factors is also assessed. Data that will be brought to bear in this chapter to prove the hypotheses will include pastoralists and farmers in Ethiopia and Kenya (at the micro level of analysis). In terms of generalizing the findings of this chapter to the wider world, since the Horn of Africa is a worst-case scenario of environmental degradation causing human insecurity (and conflict), it is possible that the chapter will prove that the relationship is not as strong as has been assumed. The Nile River (a macro level of analysis case) is also examined.

ENVIRONMENT AND HUMAN SECURITY AMONG PASTORALISTS AND FARMERS

The Horn of Africa features semiarid drylands (80 percent of the more than 5 million square kilometers), and pastoralists, who are 12 percent of the population, occupy 62 percent of land in the Horn and struggle to obtain sufficient water and rangeland.[197] The Horn of Africa features a cultural clash between "lowland" Islamic pastoralists and "upland" Christian farmers, which has been going on for hundreds of years and has centered on control over land and wealth. In Ethiopia, Somali, Oromo, and other, pastoralist ethnic groups clash with "upland" Orthodox Christian farmers from Amharic and Tigrayan ethnic groups, who have traditionally controlled state power. In Kenya, Muslim Somalis and other pastoralist groups (such as the Maasai) clash with farmers, especially from the Christian Kikuyu group (who are the ruling elite).[198]

The larger Horn refers to countries that have close relations with or are rivals of the core states, especially Sudan and Kenya and, to a lesser extent, Uganda. Sudan is especially important because of its rivalry for the past century and a half with Ethiopia. Sudan features a core group of Arab-speaking Muslim farmers from the banks of the Nile and surrounding areas who have struggled to control (often with force) vast outlying sections of the country composed of mostly pastoralists, many of whom are non-Arab speakers and non-Muslims. In Darfur, pastoralists have attacked farmers and driven them off their land as part of ethnic cleansing and genocide.

Horn states came together to create the Intergovernmental Authority on Drought and Development (IGADD) in the mid-1980s to deal with famines, drought, and other disasters which were afflicting farmers and pastoralists in the region.[199] In the mid-1990s, IGADD became the Intergovernmental Authority on Development (IGAD) and authorized the development of an early warning system to prevent or stop environmental degradation and conflict and improve human security.[200]

Environmental degradation in the Horn of Africa has been pronounced and has given rise to periodic famine and contributed to state failure and instability. Global warming has become a significant factor in the Horn; it is altering rainfall patterns, accelerating drought and desertification in places like Darfur, and leading to greater flooding.[201] The cycles of droughts and floods threaten food security and are making more pastoralists and farmers dependent on food aid. Aid dependency takes pastoralists and farmers out of the production cycle, where they risk losing their craft, and prevents them from receiving assistance to improve their production practices. The number of families without livestock or farming land is growing rapidly. This is creating a potential human security crisis.

The interrelationship between famine and conflict has been devastating in the Horn of Africa. Already noted has been the close relationship between famine, conflict, and the undermining of regime legitimacy in Ethiopia in 1973–1974 and 1984–1985 as well as in Somalia in 1991–1993. Drought and dependence on foreign aid as well as corruption undermine sustainability and legitimacy. Global warming is affecting the region, but direct evidence of warming causing conflict is difficult to confirm. For example, a study of the Turkana in northern Kenya did not find a direct link.[202] Conflicts like Darfur, between pastoralists and farmers over land, are primarily development failures and secondarily meteorological phenomena. Horn populations have been adept at dealing with climatic fluctuation for hundreds of years.[203]

In Sudan, there has been an estimated fifty- to two-hundred-kilometer southward shift of the boundary between semidesert and desert since rainfall and vegetation records were first maintained in the 1930s.[204] "Savannah on sand" represents a quarter of agricultural land and is at risk of further desertification, which is forecast to lead to a significant drop (approximately 20 percent) in food production.[205] A quadrupling of livestock numbers—from less than 30 million in 1961 to more than 130 million in 2004—has resulted in widespread degradation of range-lands.[206] Inadequate rural land tenure is an underlying cause of many environmental problems and a major obstacle to sustainable land use, as farmers have little incentive to invest in and protect natural resources. Deforestation in Sudan is estimated to be occurring at a rate of close to 1 percent per annum.[207] Population displacement, lack of governance, conflict-related resource exploitation, and underinvestment in sustainable development all produce sustainability and stabilization challenges. There are 5 million internally displaced persons (IDPs) and international refugees (Sudan has the largest population of IDPs). Competition over oil and gas reserves, the Nile River waters, and timber, as well as land-use issues related to agricultural land, are factors in the instigation and perpetuation of conflict in Sudan. Confrontations over rangeland and rain-fed agricultural land in the drier parts of the country demonstrate the connection between natural resource scarcity and conflict. In northern Darfur, high population growth, environmental stress, land degradation, and desertification have created the conditions for conflicts, which have been sustained by political, tribal, or ethnic differences. This is an example of the social breakdown that can result from ecological collapse.[208]

For pastoralists, a major problem is that the same pastoralist practices are being maintained as the population of people and animals continues to grow. The land and water cannot sustain such growth. Thus, pastoralists are becoming increasingly alienated, and it is uncertain what will they do.[209] Pastoralists in Somalia, Sudan, and the lowlands of Ethiopia, Kenya, and Uganda, are particularly susceptible to the negative effects of environmental degradation.[210] The deterioration of old clan-based pastoral orders has also brought degradation. Population growth and animal herd expansion as well as the arrogation of land for irrigation, parks, and other purposes have put pressure on grazing land. The lack of water has added to the stress on herds. Pastoralists are more dependent on markets than farmers are, and there are fewer markets for pastoralists. There is a lack of banking and credit facilities for traders. The inadequate availability of livestock marketing leads to overstocking of animals and drylands erosion. There are inflexible disease control measures that limit marketing by the frequent imposition of quarantines. Animal extension and health services are in short supply, and the capacity to limit the effects of disease is lacking.[211] Some experts have questioned whether or not pastoralism will be able to survive in the Horn.[212] Many pastoralists are dropping out of the ages-old activity due to problems of sustainability.[213]

Pastoralism is a complex activity, with complicated, well-defined rules devised and implemented by clans who know where to go in wet and dry seasons and who do not visit certain areas for a period of time to allow them to recover from grazing. The "carrying capacity" of grazing land does not apply to migrating herds. Therefore, if pastoralists are forced to switch to ranching, yields will decrease and suffering will increase.[214]

Among pastoralists, sources of conflict include scarce resources, such as water and grazing land, particularly during times of extreme hardship.[215] Desertification (caused in part by climate change) has further contributed to conflict, mainly between neighboring ethnic groups in pastoral areas. This conflict often crosses borders due to cattle raiding. Previous analysis has discussed the underlying reasons for conflict, including a lack of infrastructure to support pastoral livelihood.[216] In addition, promoting sedentary agriculture can cause alienation among pastoralists who are being forced to give up a generations-old lifestyle.

In Somalia, and to a lesser extent, in Sudan, it could be said that sustainability, stabilization, and human security challenges have created dissatisfaction among Muslim pastoralists and fostered support for Islamic extremism and terrorism.

In Ethiopia, efforts to "sedentarize" pastoralists, converting them into farmers, may not be sustainable in semiarid zones. Furthermore, the Ethiopian government is finally recognizing that it must take into account the interests of pastoralists who need to have access to grazing land and water sources and that it cannot continue to resettle farmers to cultivate pastoral areas.[217]

In Ethiopia, desertification and declining grazing land have led to the impoverishment of Somali and Oromo pastoralists, disaffection, and the declining legitimacy of the state. In areas with large clans, there is plenty of conflict over land and resources and strong and continuing ethnic tensions. In the area where Somalis and Oromos border each other, there is conflict and communal clashes.[218]

Some disaffected Somali pastoralists in Ethiopia have supported continued destabilization of the Ogaden by the Ogaden National Liberation Front (ONLF) against Ethiopian security forces. The sustainability and stabilization crisis in the Ogaden could open the door to safe havens for Islamic extremists and could conceivably generate recruits. The same could be said of some Oromo pastoralists and their support for the ONLF.

Conflict appears to be decreasing in northeastern Kenya, but structural problems remain. With good rains, there is less conflict over natural resources. There have been drought and famine in recent years.[219] Drought has caused spikes in pastoralist unrest. Pastoralists have restocked their livestock after drought by raiding other livestock from farming areas to the south. Somali pastoralists in Kenya identify more with Somalia than with their country of residence.[220] Historical neglect at the pastoral level, compounded by a lack of understanding by elites, has led to conflict. The growing practice of chewing "khat" has run counter to the encouragement of productive work. Pastoralists have been dropping out of pastoral activities due to a lack of opportunities.

In Kenya, the government has traditionally neglected pastoralists until drought and famine strike. A semiarid-lands policy should have been passed by parliament and become law but has not. There is a World Bank arid-lands project in the Kenyan president's office, but it has not been funded sufficiently to bring about the fundamental changes that are necessary.[221]

The Kenyan government has attempted to sedentarize pastoralists in large-scale agricultural schemes. On the Tano River, three kilometers on each side have been declared off-limits to pastoralists and have been given to sedentary Kikuyu farmers. Over the past thirty to forty years, land rights and tenure issues have been problematic for pastoralists. There were attempts in the 1970s and 1980s to address these issues via group ranches, which had limited success. Now there are plans to privatize, with shares sold off. District commissioners interpret the policy as they will, with instances of elitism and corruption.[222] In Kenya, land titling is unpopular with pastoralists and has led to a groundswell of dissent. In the meantime, damage to government-pastoralist relations has already been done.[223] On the Kenya-Ethiopia border (around the town of Moyale), local political figures have helped to create new political divisions among pastoralists along ethnic lines. Previously, several generations had lived peacefully together and intermarried.[224]

Horn of Africa farmers (concentrated in the uplands of Ethiopia, Kenya, and Uganda and along rivers in Sudan and Somalia) are susceptible to environmental degradation. Among farmers in the region, there are land-tenure challenges, specifically, a failure to defend customary tenure, which leads to land misuse and degradation. There are underdeveloped extension services, and consequently, poor agricultural practices continue.

In the Ethiopian (Abyssinian) highlands, the breakdown of feudal-based land-terracing in the nineteenth century, population growth, and failure to adopt new agricultural techniques have produced soil erosion on smaller plots.[225] Occasionally, environmental degradation has contributed to famine.[226] Tree cover in Ethiopia is only 2 to 3 percent, which helps explain extensive erosion.[227] In Ethiopia, government policy has been to keep people on the land. Farmers do not want to leave the land because all land is government-owned, and land that is vacated will be taken and given to other farmers. The murder rate over land issues is very high.[228]

Research on the consequences of moving farmers from northern Ethiopia to the south in the wake of the 1980s famine has found that grain farmers from the north imported agricultural practices that caused soil erosion and environmental degradation. Once in the south, the harmful practices were adopted by indigenous farmers, who sought to emulate the immigrants and become more prosperous.[229] In Kenya (and other parts of the Horn), there is overcrowding of land and environmental degradation that leads to human insecurity.[230]

PROMOTING ENVIRONMENTAL SUSTAINABILITY AND THE IMPACT ON HUMAN SECURITY

Given the preceding analysis, the forthcoming part examines ways of overcoming environmental degradation and promoting human security. At issue is the extent to which human security will improve if environmental sustainability is enhanced. Also to be considered is the possibility that improving the environment and human security will lessen violence and conflict and support for extremism. It can be assumed that overcoming environmental degradation and enhancing sustainability and human security is a complex challenge and requires a variety of approaches, including a long-term structural approach.

Globally, the process of slowing climate change depends on the reduction of greenhouse gases. Reductions are especially needed in the United States and other developed countries and in rapidly industrializing India and China. In the Horn of Africa, there is an undeniable trend towards climate change that is now being addressed. Vulnerability to climate change is being factored into the policymaking of a number of governments. Reductions in greenhouse gases can be made through rural electrification and other means of producing energy than wood fires and charcoal.[231] Alternative energy sources, such as solar power and geothermal energy, especially in the Horn, have the potential to reduce greenhouse gas emissions. However, there are questions of capacity, specifically, the ability of African governments to frame the issues and to obtain the necessary training to tackle climate change. At the moment, energy issues are not dealt with sufficiently, and aid programs that focus on alternative energy sources are lacking.[232]

The reduction of population growth requires the education and empowerment of women and shifting from small-scale farming and pastoralism to other economic activities, such as small-scale enterprises.[233]

More development projects are needed linking population, development, and the environment. Women need to be assisted in the process of overcoming subjugation and the obligation to have as many children as possible. Rural electrification could help women and girls to overcome the task of gathering firewood and provide more time for education and employment.[234]

Overcoming desertification entails the introduction and inculcation of sustainable farming and pastoral practices, tree planting, and water management, which in turn require the development of effective extension services. Preventing famine depends on developing early warning mechanisms so that governments and the international community can react to mitigate the effects of severe drought, flooding, and man-made resource shortages.

Enhancing sustainability and human security involves interventions on a number of fronts that reduce conflict over resources and build the capacity and increase the legitimacy of the state in the eyes of its citizens. The introduction of more sustainable farming and animal management methods as well as markets and roads is one way to increase legitimacy but requires a considerable commitment of time and resources by outside agencies, as the capacity of the state as well as the private sector are developed.[235] Water management is another method; in the short term, the drilling of boreholes and the building of small dams can enhance sustainability and win support from the local population. In the longer term, strategies for water conservation and management as well as government capacity to manage water need to be developed. These interventions over a sustained period of time can begin the process of connecting pastoralists and farmers to government, thereby increasing legitimacy.[236] At the micro level, development projects for pastoralists and farmers, the education and employment of women, and microfinance as well as tree planting can enhance sustainable development.[237]

The Kenyan government has set up a food security working group, and donors are investing via the government. However, the problem of food security has been tackled in natural resources management due to a lack of awareness that social safety nets are needed to improve livelihoods, entrepreneurial activities require greater support, and better land management requires incentives.[238]

In southern Sudan, "environmental governance" has just started, but the impact of the various options for shifting back from aid dependence to autonomous and sustainable livelihoods has been examined. Early signs are the provision of food and other emergency aid to some 15 percent of the population.[239] National and regional governments are assuming increasing responsibility for investment in the environment and sustainable development. All UN relief and development projects in Sudan have integrated environmental considerations in order to improve effectiveness. Investment in environmental management is needed to support lasting peace in Darfur and to avoid local conflict over natural resources elsewhere in Sudan. Capacity must be built at all levels of government, and legislation must be improved to ensure that reconstruction and economic development do not intensify environmental pressures and threaten livelihoods.[240]

In the Horn of Africa, in order to promote sustainable development and human security for those remaining as herders, pastoralists must be provided with the means to maintain their herds and market their animals. Those who cannot make it as herders need to be provided with alternative means of making a living. The provision of pastoralists with more clearly recognized legal tenure over their resources will lead to better land use practices. To establish their rights, communities require a stronger political voice. A participatory approach to project development and implementation should be promoted, and local pastoralist institutions such as trading associations and peace committees should be built. Pastoral self-governance should be strengthened.[241] Also, pastoralists who continue to maintain herds need to diversify their income sources.[242] There must be changes in mind-sets and opinions regarding pastoralists and marginalized areas in order to change policy.[243]

Pastoralists are demanding services, including markets, abattoirs, and cold storage as well as market information and microfinance. In particular, markets are in high demand by pastoralists. In Kenya, a holding pen was installed as a safe place for livestock that would induce Samburu pastoralists to take the risk of walking their animals eight days to market. Technology could be brought to bear through inexpensive radios, cell phones, and other communications to provide information for pastoralists to inform them when the price is right to sell their livestock. Cash-based insurance policies for herds should be considered. Programs should be supported that aim to reduce short- and long-term vulnerability throughout the drought cycle. A livestock health certification system is needed to satisfy the demands of the importing countries. Livestock marketing cannot be considered separately from livestock disease control and conflict, and all three areas have technical, infrastructural, social, economic, and political and policy components.

More pastoralists will be positively affected if the emphasis is on livelihoods rather than on simple commodity production. Pastoralist society cannot survive with land degradation, and long-term solutions are needed. Pastoralists need education and exposure in order to provide alternative lifestyle approaches. Mobile schools and clinics, which have worked well elsewhere, hold great potential for pastoralists in the Horn of Africa. In particular, women's literacy programs curb population growth. There is a need to develop rural financial markets.[244]

In Kenya, a participatory approach has been developed to project development and implementation, and local pastoralist institutions, such as trading associations and peace committees, are being built. In order for the government to provide police, teachers, and health workers to Somali pastoralist areas, they must be attracted to move with decent infrastructure, roads, water, housing, and electricity. Also, a national service program for Kenyan graduates to work in pastoralist areas would be helpful.[245]

In Kenya and elsewhere, a solution for pastoralists is civil society–driven local government recognized by central government as having mandates. This will build bridges across the divides. Consensus building among representational groups is important in conflict resolution in Somali areas and elsewhere. Marketing and other services should be provided in Garissa for pastoralists to sell livestock and develop entrepreneurial activities. Good roads are needed to ship cattle to Mombasa and Djibouti.[246]

In Kenya, in regard to environment, human security, and extremism in Somali pastoral areas, Somalis are not strict Sunni Muslims, and moderate forces exist. As a group, they are not susceptible to extremist philosophy, but if forced to choose sides, they will go with the Islamists. The key to encouraging moderation is to provide economic growth equitably and engage with the local population.[247]

The Kenyan government requires a mind-set change in relation to Somalis and other politically marginalized groups in order to move towards sustainability and human security. The government has tended to take a divisive approach, making issues political as well as resource-based. Pastoralists require assistance in managing the excessive growth of population and animals. Ways must be found for herders to move to other pasture and water areas so that conflict can be avoided with sedentary agriculturalists. Visual tools are being developed to assist pastoralists, including tools that show ethnic overlay, trade routes, and markets, and enable them to deal with resource and pasture-access issues. Access to the political process is essential to provide pastoralists with voice and participation at the national and provincial levels.

Fifty percent of the GDP in Kenyan agriculture comes from pastoral activities throughout the country. The marginalization of pastoralists is partly due to misperceptions. Marginalization is now being overcome now by technology (e.g., cell phone towers and access in the rural areas); pastoralists now have access to market information and can engage in rural banking and develop livestock sales. Funds can go into other entrepreneurial areas besides livestock. Therefore, the best initiative for sustainability, stabilization, and human security is providing access to economic resources and economic development.[248]

The Ethiopian government is still searching for answers regarding pastoralists in the lowlands.[249] Restrictions have been placed on grazing in the lowlands, resulting in revegetation. However, with ten to eleven months of drought and only one month's rain, animals have to keep moving to find pastures. The alternative is to establish big ranches, which require the reorganization of the land-tenure system. Grazing in the driest period of year along riverbeds is being managed. The Ethiopian Environmental

Protection Agency wants to expand crop cultivation along riverbeds and beyond with irrigation and pasture, but the riverbeds have to be maintained, which is a challenge. Irrigation is expanding to arid areas of Afar and the Ogaden. Pastoralists are being settled as farmers, but it is challenging, as they need to have complete awareness of the crop cultivation cycle.

In southern lowland Ethiopia, where there is increasing frequency of drought, a USAID-assisted pastoral livelihood initiative has involved destocking during the early phase of drought and keeping herds alive during drought, with enough livestock kept to retain a herd's viability. This is making pastoralists more market-oriented. Chilled meat (goat, mutton, and beef) is being shipped to Middle East markets. The live animal trade through Djibouti and Saudi Arabia has provided access to markets and has been critical for pastoralist survival.[250]

In regard to sedentary farmers in the Ethiopian highlands, the government is implementing a land certification and registration process, which is the first step towards land reform and curbing degradation.[251] In the highlands, small-scale dams, biogas, and rural electrification are being developed to prevent greater land degradation.[252] There is a push for indigenous, fruit-producing, and fodder trees to be planted in the highlands.[253] The government is trying to help farmers to produce oil seeds, horticultural crops, leather skins, and lentils for the export market.

In the Ethiopian highlands, there is a sustainability emphasis on organic agriculture. The traditional system of free-range grazing has now been curbed to assist in the intensification of organic agriculture through compost development. Conversely, bringing food to animals allows revegetation to occur, and soil erosion has been reduced.[254]

The literacy rate of school-age children has risen from 46 percent in 2000 to 90 percent today, thanks to a crash literacy program. This helps with agricultural extension, which has become widely present. However, mountainous terrain and intensive crop production prevent "green revolution" formulas of advanced seeds and fertilizer from working because the climate is not uniform. Instead of agrochemical fertilizers, promoting the use of compost and training farmers to develop the skills to use compost will retain moisture levels in soils. Ethiopia is emphasizing organic agriculture, with no genetically modified crops and with smallholders making their own compost. In Ethiopia, crop genetic diversity has been maintained, which makes pest control easier. The worldwide demand for certified organic exports is increasing, and Ethiopia is exporting "boutique crops."[255]

Rural electrification and telecommunications development are progressing, facilitating Internet access in each village as well as lighting, electrical appliances, radios, and televisions. In the next decade, if the current rate continues, most people in rural settled areas will have electricity. Potential for hydropower is great, with a focus on electrification. There is good potential for geothermal power generation (not solar power or wind power). Ethiopia will be exporting electricity to surrounding countries. There are currently two hydropower generation projects. In one year's time, one project is expected to increase hydropower by 40 percent. Another project in four years will increase electricity output further.

In the highlands, community empowerment and devolution is the key to coping with expanding population and soil erosion. Local communities must have control over their own local resources and environment. They are the ones who can stop free-range grazing. Federalism is also important. Provincial governments are passing laws that do not contradict central government laws. This practice started in Tigre Province, where the provincial government started a land recovery program. Tigre was a good place to start because there was a stable situation (the TPLF has long been the governing force) and a severely degraded environment, where improvement would provide a demonstration effect. The regional agriculture bureau became interested in the project. In three years, the land recovery approach was adopted in the agricultural extension system throughout Ethiopia.[256] Ethiopia has an early warning system, developed after the 1984–1985 famine. Rain gauges have been distributed throughout the country to tell when rainfall totals are down. Monthly reports are provided. District governments report immediately when unforeseen events (e.g., drought, flooding, or locusts) occur.[257]

Partners in Sustainable Development and Human Security

UN agencies, donor agencies (such as USAID), and NGOs are partnering with pastoralists to achieve sustainable development and human security. There have been changes in the UN system to give UNDP and the UNDP resident representative greater authority. A problem is getting humanitarian relief agencies to think more long term (as development organizations do).[258] An example of partnership among donors and NGOs has been the Regional Enhanced Livelihoods in Pastoral Areas (RELPA) program, managed by the Nairobi regional office of USAID, which is promoting sustainability and stabilization in the region.[259] With the growth of population and livestock, RELPA is assisting people, especially women, who want to leave pastoralism to find other livelihoods. Savings clubs have been supported by NGOs but have also been spontaneously supported by women. Another RELPA focus group is young men who are looking for alternative livelihoods.

There are projects that are promoting sustainability and stabilization in pastoralist areas with a number of NGO partners.[260] Some NGOs work for conflict mitigation and provide conflict resolution training to deal with pastoralist water disputes and subclan conflicts.[261]

A joint-programs approach is being developed to promote sustainability and human security. Such an approach is regional, integrated, and comprehensive, which is essential because pastoralism is regional and boundaries do not constrain groups. The task is to make steps forward. Thus far, there have been procurement issues and delays. The program needs more than two years' funding to be effective. Also, livelihood interventions have not had the impact desired, as they have not addressed or understood underlying conflict dynamics among pastoralists. In any case, it is unlikely that "alternative livelihoods" programs can mitigate support for Islamic extremism.[262]

A consortium of groups has formulated a cross-sectoral program across the Ethiopian and Kenyan borders. Case studies in Ethiopia and Kenya indicate that there is now better reporting response regarding conflict and early warning and drought (this is difficult to do in Somalia).

The Integrated River Basin Management Project works with the private sector. There is a need for resources in order to provide access to safe water. In addressing issues of education and health, mobile schools and clinics have been proposed in pastoral areas. The health issue is vital to address. A quarter of the pastoralist population has acute malnutrition, compounded by lack of stable health service and lack of hygiene.[263] A regional conflict-mitigation group has opened a market in Mandera near the Kenya-Ethiopia border. The Kenyan government wants to link with the military-diplomatic initiative in the Mandera triangle, which is being developed as a US "3D" (development, diplomacy, and defense) area.[264]

THE IGAD-CEWARN EARLY WARNING MECHANISM

IGAD-CEWARN was established in 2001 and is providing early warning concerning environmental degradation, drought, famine, and conflict. The strength of CEWARN is the state-of-the-art software. There are fifty-two sets of selected indicators of communal variance, areas reported, media reporting on conflict, and environmental context. Field monitors provide weekly reports on specific incidents as they happen. Information flows to the national level and to IGAD-CEWARN. The weakness is lack of action by governments. The well-established response mechanism needs to be programmatically designed and developed. Another problem is the diverse source of funds (60 percent from USAID, 30 percent from German GTZ, and only 10 percent from member states).[265]

The Ugandan government has used daily reports from IGAD-CEWARN to deal with problems of drugs, arms, and human trafficking by pastoralists. The Ugandan government adopted a disarmament strategy with NGO funding. The Ugandan army was used to disarm the pastoralists. Many community members (especially the Karamojong, in the Karamoja region died as a result.[266] The IGAD-CEWARN approach is being made to work at early warning and conflict management among pastoralists at a lower level. Then, it can be developed to manage bigger conflicts and disasters involving states.[267]

CJTF-HOA AND PROMOTING SUSTAINABILITY, HUMAN SECURITY, AND STABILIZATION

The Combined Joint Task Force-Horn of Africa (CJTF-HOA) has been reaching out to pastoralists in the primarily Somali regions of Ethiopia and Kenya—with six civil affairs companies and five force protection companies. Civil affairs companies have developed water, school, and health clinic projects in strategic locations on transit routes from Ethiopia and Kenya into Somalia. The projects have gained favorable reviews from Somalis, and the news was passed to Somalis inside Somalia in the Juba River Valley and other parts of the country.

CJTF-HOA aims to promote sustainable development, human security, and stabilization together. CJTF-HOA claims to have contributed to overcoming ethnic and religious hatred that can lead to terrorism.[268] At issue is whether or not it has achieved the aim of making Somalis in Ethiopia and Kenya want to be part of their respective states. No?

An example of sustainable development has been water projects that can be managed by a coalition of militaries, international and national development agencies, and NGOs. The CJTF-HOA Gode water project in the Somali Ogaden region of Ethiopia serves as an example. Water-drilling equipment was flown in by two C-17 and seven C-130 flights. An important lesson was to listen to the locals. The water at Gode was not potable for humans, but animals could drink it and accumulate minerals. The locals showed the civil affairs company where to drill and how to avoid conflict among the clans.[269]

CJTF-HOA is cooperating with U.S. embassies and USAID in Nairobi and Addis Ababa in developing the three "Ds"—diplomacy, development, and defense—to work together to promote sustainable development and win the hearts and minds of pastoralists in the Horn of Africa.[270] The CJTF-HOA experience demonstrates the importance of culture and language awareness, and Somali culture and language awareness is part of the operation. It has the potential to reach out to other ethnic groups in the region.[271]

The "3D" concept (diplomacy, development, and defense) is inculcated during four months of training. In Kenya, coordination between CJTF-HOA and the U.S. Department of State and USAID is developing. The goal is theater security cooperation, specifically, promoting peace and stability, preventing conflict, and preventing terrorism.

CJTF-HOA teams have been building schools and clinics and developing capacity in marginalized areas. It is important to understand the region, local leadership and population, clerics, and store owners. Also important is to know CENTCOM and CJTF-HOA commanders' intent. Picking projects involves bimonthly meetings at the U.S. Embassy, Nairobi, with representatives from the U.S. Department of State, USAID, CJTF-HOA, and civil affairs liaisons (all stakeholders are represented). Ninety percent of the proposed projects are approved. CJTF-HOA representatives consult with the Kenyan government and local Somalis. CJTF-HOA teams try to imbed Kenya Ministry of Defense personnel with CJTF-HOA personnel.[272]

CJTF-HOA civil affairs teams have connections with NGOs via USAID. The NGOs are aware of CJTF-HOA water projects. One cultural problem arose during a medical exercise (MEDCAP) in Garissa in which people were provided vitamin packets that also contained condoms, which are taboo. This reflected a lack of cultural-awareness training on the part of CJTF-HOA personnel. Before teams arrived, the Somalis were very anti-American. Now, teams have had an impact through interaction with local people. There is more positive opinion of the U.S. military and the United States in general. There is greater acceptance by the imam and the local commissioner.[273]

In regard to strategic communications, CJTF-HOA has been booking time on Somali-language radio stations. However, there is little or no measurement of the winning of hearts and minds at CJTF-HOA headquarters in Djibouti. In terms of measuring effects, the concern is how many terrorists are killed or captured. A long-term presence is needed to build trust and build links between Somalis and Swahilis and the Kenyan government.[274]

The direct impact of CJTF-HOA civil affairs activities in improving relations between pastoralists and the Kenyan and U.S. governments remains unknown. At issue is long-term sustainability of CJTF-HOA interventions. The problem is that the basic goal of the Ethiopian and Kenyan governments is to prevent their Somalis from having anything to do with a "greater Somalia." The best way to do that is to keep the Somali population quiet and at bay. CJTF-HOA should be careful in trying to win hearts and minds; it could create jealousy among the Kenyan and Ethiopian officials. The Kenyan government does not want civil affairs activities to work too well to help Somalis. For example, when CJTF-HOA drilled a borehole for Somali pastoralists, the population grew around the facility, which created disputes over whose cattle should be able to drink and who should maintain the well. A borehole creates a permanent settlement, which creates demand for schools, housing, and clinics. Therefore, it is important for USAID and the U.S. Department of State to establish partnership with Somalis as well as Kenyan and Ethiopian officials to guarantee long-term sustainability.[275]

CJTF-HOA civil affairs teams have been challenged, especially given their short-term horizons. A recent increase from six-month to one-year rotations has made the teams more open to collaboration with U.S. State Department and USAID (3Ds). The CJTF-HOA focus has been too much on counterinsurgency. The United States should make a high-profile intervention, given the lack of trust between Somalis and the governments of Ethiopia and Kenya, in order to provide a forum in which all sides can air their differences.[276]

The "3D" dialogue has taken lots of effort and energy. The hope is that it leads somewhere constructive. The CJTF-HOA concept of building state capacity by developing infrastructure—constructing schools and clinics—is a nebulous concept. Presently, the USAID goals are to try to keep CJTF-HOA from doing harm, steer the military away from bad ideas, and develop a good spirit of cooperation over the years to come. The U.S. military has relied too much on personnel from Washington, DC, when there are reliable U.S. personnel available at local embassies. USAID is now spending less time on pastoralists and more time and funds elsewhere. Activities among Somalis have been harmed by U.S. support of Ethiopia in Somalia. The NGO reaction to CJTF-HOA and USAID interventions has been varied—some are now more flexible and willing to engage.

The CJTF-HOA program has appeared to benefit the Ogaden National Liberation Front (ONLF) rebels in the Ogaden more than the Ethiopia People's Revolutionary Democratic Front (EPRDF) regime. After the attack on April 29, 2007, on the Chinese and Ethiopian oil workers by the ONLF, the regime suspended CJTF-HOA operations in the Ogaden in order to deal with the ONLF. This leaves open the question of how to win Ogaden Somali hearts and minds so that they do not support rebels but come to associate more with the provincial and national-level governments.[277]

CJTF-HOA civil affairs teams in Garissa, Kenya, have talked to imams, businessmen, and others. There was a "paranoid response" and "negative perceptions." U.S. policies in Somalia and elsewhere remain a problem in winning hearts and minds. The United States should not have been so heavy-handed in Somalia.[278]

Critics assert that the digging of boreholes by CJTF-HOA needs to be integrated into a sustained program, which improves logistics and energy and water supply in a holistic and sustained fashion that should provide a model for CJTF-HOA.[279] The CJTF-HOA civil affairs approach is blurring lines between civilian and military. Previously, civilians have worked within the framework of traditional pastoralist norms and hospitality, and their development program goals have been appreciated by pastoralists. Military intervention has negatively changed the impact of NGO interventions. The United States needs to have a clear understanding of what it is trying to achieve in pastoralist areas and decide how programs will be implemented to change the local environment. USAID should take the lead, the U.S. Department of State will engage imams and other actors, and the Department of Defense will build schools and boreholes. Unless the dots are joined, the agencies will be working at cross-purposes, reinforcing negative impressions. The U.S. agencies will be manipulated by national governments. The 3D approach, with the differing strengths and perspectives of different partners, presents problems: "defense" has far more resources than the "diplomacy" or "development" parts of the triad. In addition, CJTF-HOA personnel need to find ways of gaining access to the expertise that NGOs possess.[280]

The Nile Basin Initiative (NBI), Sustainable Development, and Human Security

The current population of the ten Nile riparian states is close to 350 million people, with an expected growth rate that will nearly double the population to more than 600 million by the year 2025. The pressures of population growth and the rise in demand for food and water ensure that all the states will be searching for ways to gain more access to Nile waters for sustenance, transportation, and agricultural and energy production, which will cause environmental challenges and will affect human security.

In the early 1990s, a Nile Council of Water Ministers (NILE-COM) was formed to discuss river issues. In 1997 Egypt made an overture to Ethiopia to discuss Nile issues bilaterally, which led to an exchange of diplomatic notes and a proposal that led to the Nile Basin Initiative (NBI).[281] In 1998 nine of the ten Nile riparian states recognized the need to establish a cooperative arrangement to deal with competing demands on water resources and established the NBI.[282] The World Bank, UN Development Programme, Canadian International Development Agency, African Development Bank, and other agencies facilitated the formation of NBI and have continued to encourage its development. To facilitate lending to NBI projects, the World Bank fostered the formation of two subregions,[283] the Eastern Nile and Nile Equatorial Lakes. The initiative was launched to "create a regional partnership to facilitate the common pursuit of sustainable development and management of Nile resources."[284] The NBI consists of the NILE-COM, Nile Technical Advisory Committee (NILE-TAC), and Nile Secretariat (NILE-SEC). The three institutions are supposed to jointly interact to create a coordination mechanism and an enabling environment to realize their shared vision through action on the ground. The NBI is a broad-based, basinwide program of collaborative action, exchange of experience, and capacity building to ensure a strong foundation for regional cooperation.[285]

The ultimate effectiveness of the NBI is in doubt, however. One commentator assesses the NBI as follows:

> For the first time in history, all the Nile riparian states have expressed their commitment to a joint initiative. However, expressing concern over common freshwater resources and taking concrete action are two different things. Though Egypt is talking about basin-wide cooperation, it continues to develop massive new water projects unilaterally within its borders. Furthermore, there has been no recent reduction of its dependence on the Nile water. On the contrary, Egypt's demand for water is increasing considerably.[286]

Egypt's hegemonic behavior has been an obstacle to establishing a cooperative Nile regime. Egypt has been accustomed to using the Nile as it pleases and rejecting other states' plans to use Nile waters. The NBI offers a framework for Egypt to change its behavior and begin more cooperative conduct. However, a collective action problem exists: Egypt as the hegemon is not willing to provide a plan, backed by its resources, for sharing Nile waters that would attract other NBI states and spur cooperation.[287]

Several states, such as Kenya, Tanzania, Rwanda, Burundi, and the Democratic Republic of the Congo, have been generally indifferent to the plight of the Nile basin and need to be convinced of the importance of sharing and managing its waters. The difficult relationship between Egypt and Ethiopia, fueled by nationalism, poses a deeper problem for the NBI.

Although the NBI provides a useful framework for cooperation, the large number of actors and interests means that

> the Nile Basin Initiative has until now functioned based on the strategy of securing consensus of all ten riparian countries on less controversial issues, while postponing more difficult ones. By failing to address the core issues and projecting a superficial cooperation involving a larger number of actors, the NBI is very likely to fail.[288]

One way forward is for interested states (e.g., Egypt, Sudan, Ethiopia, and Uganda) to forge agreements among fewer actors. Uganda is helping to revive the Kagera Basin Organization with Rwanda, Burundi, and

Tanzania and could lead in forging a Lake Victoria agreement with Kenya and Tanzania. Egypt, Sudan, and Ethiopia could come to an agreement on the development of energy and irrigation projects on the Blue Nile. Sudan and Ethiopia are in the process of jointly developing a "border dam" for hydroelectric power and irrigation, though Egypt might eventually veto such a project.[289]

Another avenue would be to elicit greater involvement of the international community. The Nile riparian states do not possess the capital or technical competence needed to develop projects that will have regional benefits. Each state, while espousing regional cooperation, still acts within its own self-interest. Nonpartisan third parties can play a role in resolving disputes and encouraging cooperation and development. Developed states have not invested considerable time and energy in assisting the NBI, choosing to monitor the situation from afar and through membership in the World Bank, UNDP, and other multilateral agencies. Advanced states can best help the situation by providing financial assistance where necessary, providing arbitrators to help adjudicate disputes, and remaining active behind the scenes, goading parties into action where necessary and acting as a restraint on others when required. Only through global involvement and local cooperation on river development will the ten countries sharing the Nile resources be able to contribute to the NBI's vision for water management.

The eight states that did not sign the Nile waters agreements of 1929 and 1959 concur that they are not obliged to abide by the provisions of either treaty. The pacts were agreed upon and signed by Britain, acting on behalf of its colonies before they became independent states. Therefore, most riparian states believe the treaties are nonbinding and in need of renegotiation. Thus, the management of the Nile issue requires involvement by international organizations and the promulgation of a new treaty involving all the parties.[290] The Nile Cooperative Framework Agreement, supported by nine riparian states and opposed by Egypt and Sudan, seeks the establishment of a permanent Nile River Basin Commission through which member countries will act together to manage and develop the resources of the Nile, instead of asking the permission of Egypt and Sudan.[291]

The Kenyan government commissioned the Konrad-Adenauer Foundation to study its obligations under the Nile treaties, which found that Kenya was not obliged to adhere to standards set forth in an agreement to which it did not sign as a sovereign nation.[292] Uganda's water supply and electricity generation through the Owen Falls Dam are being affected by other states' use of waters that flow into Lake Victoria, the major source of the White Nile. Kenya is trying to tap into the waters of the lake in order to aid in its development. Because of steady drought in the region, Tanzania has proceeded with a plan to tap into lake waters to supply water to drought-stricken areas.[293] Therefore, it is in Uganda's interests to foster cooperation in the management of Nile waters.[294] In the meantime, the water level of Lake Victoria has been dropping, the water hyacinth has been choking the lake, and farms are draining several lakes.[295]

The Ethiopian position is that the country can freely tap into the water resources of the Nile.[296] Ethiopia must be prepared to defend its development of the Blue Nile, because Egypt will probably react in a hostile manner. Ethiopians see the United States and the European Union backing Egypt in any dispute over the Nile.[297] During the 1970s and 1980s, devastating droughts in the region claimed more than a million Ethiopians from famine, partly because the country had no means of capturing the Nile waters for irrigation and drinking.[298] Ethiopia is not yet able to dramatically expand its use of the Nile in the near future. Already the most populated country among the Nile riparian states, estimates indicate that Ethiopia's population will climb from 68 million people today to 127 million by 2025.[299] Ethiopia continues to underutilize its arable land because it lacks irrigation capabilities. A 2000 African Development Bank study showed that Ethiopia had only irrigated 190,000 acres of farmland out of a potential 3.6 million acres (5.4 percent).[300] Electricity is available to less than 10 percent of the people, and, for those who have it, the cost is so high that it must be rationed.[301] Furthermore, farmers are too scattered for the government to effectively deliver electricity. Estimates indicate that only about 3 percent of the 110 billion cubic meters of freshwater that originates in Ethiopia is utilized in the country.[302]

As Ethiopians seek to improve their agricultural production, socioeconomic conditions, and human security over the coming decades, they will do so partly by harnessing the country's Blue Nile resources. Ethiopia is currently forging ahead to develop the irrigation and energy potential. It wants to dam some of its smaller rivers and tributaries to the north of Lake Tana in order to capture water for irrigation and electricity generation, though greater use of the Blue Nile remains a long-term project.[303] Recently, Ethiopia has announced plans to build five more dams in the Lake Tana region.[304]

With the end of the north-south Sudanese civil war, opportunities for development of the Nile have emerged. Sudan is currently planning on building a dam on the Blue Nile near its border with Ethiopia to generate hydroelectric power, and the country could dramatically expand its uses of the Nile for irrigation and other agricultural purposes.[305] The Sudd marshlands remain a major issue between the north and south Sudan. The Jonglei Canal, which would provide greater navigability on the White Nile, would dry up the marshlands, which constitute an important ecosystem for southerners. The Chinese and Egyptians are supporting Khartoum in efforts to revive the canal, which could have a devastating impact on human security in southern Sudan.[306]

The preceding analysis of struggles over the Nile River demonstrates that "structural scarcity," created by Egypt's claim to veto power over the use of Nile water, is hampering upstream states and peoples from overcoming water shortages, from developing hydropower as an alternative to wood and charcoal fires, and from building irrigation systems that would enhance human security. In the future, demand- and supply-induced scarcity of Nile water will grow as upstream states and peoples more fully utilize the river.

CONCLUSION

The severity and extent of human insecurity and the consequences of environmental degradation are as great in the Horn of Africa as in any other region. Drought, water shortages, and desertification have produced famine and other forms of human insecurity as well as weakened states, conflict, and support for Islamic extremism and terrorism. Environmental degradation and resource scarcity have been largely demand-driven, especially among growing populations of land- and water-hungry pastoralists and farmers. Famine has been the result of the interplay of growing populations, environmental degradation, and drought as well as conflict and states' lack of capacity to react. In particular, the predicament of Horn pastoralists is serious and seems destined to grow worse, given growing populations, desertification, and global warming.

Horn governments, donors (including USAID), UN agencies, and NGOs have devised a range of policies and programs, including food security, early warning, and integrated population, health, and environment programs, which have slowed but not reversed environmental degradation and human insecurity. Greater state capacity and local participation are needed to achieve sustainable development and human security. U.S. efforts to bring sustainable development and human security to pastoralists in the Horn have the larger aim of winning hearts and minds in order to prevent support for Islamic extremism and terrorism; the success of these efforts is to be determined. The fundamental divide between Horn pastoralists and their governments is a barrier to progress that will not be easily overcome.

The structural scarcity of Nile River water, a result of Egypt's insistence on a monopoly, has deterred riverine states and peoples from utilizing the water for irrigation and hydroelectric power. The Nile Basin Initiative and Nile River Commission can help to build cooperation, sustainable development, and human security. However, the United States and other powers need to become more engaged to persuade Egypt and Sudan to work with other riverine states.

In sum, awareness of the problem of environmental degradation and deteriorating human security is growing, as is international action to ameliorate negative trends. However, environmental degradation continues in the Horn of Africa and will contribute to deteriorating human security and conflict for some time to come.

CHAPTER 3

BITTER ROOTS

THE OBSTACLES TO PEACE
IN THE CASAMANCE CONFLICT

Mark Deets[307]

INTRODUCTION: NO WAR, NO PEACE

It really was quite a shame. The refugees had begun to return to their land and to their villages. NGOs planned to start pulling land mines out of the fields. The government had begun building a new road in hopes of reviving the almost-extinct tourist sector. Perhaps the road would revive agriculture and the fishing industry as well. It seemed as if all was falling into place in early 2006 to finally end West Africa's longest-running civil conflict—the Casamance conflict in southern Senegal. Two of the three warlords that made up Attika, the armed wing of the Movement of Democratic Forces of the Casamance (MFDC), had agreed to lay down their weapons, come out of the forest, and give peace a chance. But Salif Sadio was the recalcitrant hard-liner who refused to approach the bargaining table with the Senegalese government. Therefore, with Senegalese military logistical and intelligence assistance, the army of Guinea-Bissau had cornered Sadio and a few hundred of his fighters just south of the Senegal-Guinea-Bissau border. However, Salif was not about to come out of the forest and give up his weapons, and so the fighting began…again.

As civil wars go in West Africa, the Casamance conflict was "not that bad," but it was bad enough to waste the potential of Senegal's greenest, most bountiful region. It seemed such a shame. As the U.S. marine and defense attaché to Senegal, the Gambia, Guinea-Bissau, and Cape Verde from January 2005 to July 2007, I found myself repeatedly asking certain questions: Why had real peace in the Casamance been so difficult to achieve for twenty-five long years? Was it all Sadio's fault, as several Senegalese military officers told me? Was there something to the secret, mystical forest meetings so central to the ethnic and cultural identity of many of the *maquis* (rebels)? What caused the *maquis* to pick up the gun in the first place? Was the conflict better explained by the region's political economy or its cultural identity?

When studying the Casamance conflict to help facilitate the U.S. embassy's role in the peace process in 2006 and 2007, I discovered that the roots of the conflict ran deep. As I pulled up one root that I thought would enable me to fully understand the violence, I discovered more roots buried a little deeper—some that blended into or branched off from yet other roots. Each root added to my understanding but at the same time made the problem more complex. The more I learned, the more I realized how little I knew. After a few months on the ground in Dakar and one trip to Ziguinchor (the regional capital of the Casamance), I thought I had all the answers. I attributed the conflict to Senegal's colonial legacy and its resulting political economy. Later, I decided that at least part of the answer lay with the Jola ethnic group, the *maquis'* center of gravity. Sure, there were Mandingo, Fulani (*Peul* in French), and Balanta in the MFDC, too, but as a Casamançais journalist once told me,

"Not every rebel is a Jola, but every Jola is a rebel." Then I discovered that the Jola were not monolithic. Different kinds of Jola lived in different parts of the Casamance. Some of them supported the MFDC more than others. My aim here is not an ethnolinguistic study[308] nor to offer political solutions to the Casamance conflict. Rather, I aim to elucidate some of the drivers and offer a few potential outcomes of the conflict. Ultimately, in spite of the occasional cease-fire, the *maquis* will continue fighting until the benefit from doing so becomes less than the cost.

To support this thesis, I examine the drivers of the conflict based on a model from Robert Lloyd, a conflict resolution specialist at Pepperdine University. According to Dr. Lloyd, most conflicts have four common drivers. First, one side must have a grievance. The grievance may be a result of some perceived neglect or discrimination on the part of the other side. Second, the aggrieved group must assert an identity that distinguishes it from the "other," that facilitates groupthink along the lines of "us" versus "them." Third, the group must be able to wield power in the form of commitment or resources. Finally, the group must agree on a valid spokesperson.[309] Since the start of the Casamance conflict, these four drivers have waxed and waned with the fortunes of the MFDC, but they have always persisted. Therefore, while providing some historical background, I will show how each driver applies in the Casamance and thereby supports my thesis.

HISTORICAL BACKGROUND

There was much to cheer about the cease-fire of December 2004, a few weeks before my arrival at U.S. Embassy, Dakar. After MFDC and Senegalese representatives signed the cease-fire, the BBC had announced, "Senegal's southern rebellion starts to end"[310] and "Crowds cheer Senegal peace deal."[311] The crowds cheered because they were fed up with a conflict that had endured two decades and threatened to stretch into a third. The conflict began in December 1982, when, after a meeting in a "sacred forest," separatists marched on Ziguinchor to demand independence for the Casamance—only to be greeted by Senegalese security forces in a crackdown that left several marchers dead and more seriously injured.[312]

Since then, years of violence and instability have taken their toll. After the cease-fire, Mark Doyle of the BBC noted, "The violence in the Casamance region has rarely received much publicity. But it is estimated that roughly the same number of people have been killed in the conflict—around 3,500—as have died in the unrest in Northern Ireland."[313] Added to this carnage are the 60,000 refugees who fled the fighting into the Gambia and Guinea-Bissau, as well as the ninety-three villages confirmed to have land mines and the 149 suspected of having them, for a total of 90,000 villagers affected by land mines.[314] Since the conflict began in 1982, nearly 700,000 people have been living in "a state of insecurity."[315]

Though it had taken longer than the hundred days promised by Abdoulaye Wade before he assumed the Senegalese presidency in 2000, many Senegalese, including most Casamançais, felt elated by the cease-fire, viewed as the first step to a real peace. After false starts in 1991 (the first cease-fire), 1999 (Banjul I), and 2000 (Banjul II), the Senegalese hoped this deal would finally bring peace and prosperity to the greenest, most resource-abundant, and most naturally beautiful region of Senegal, returning the tourists who abandoned the white beaches of the Casamance due to the conflict in favor of the brown sand and less beautiful—but peaceful—beaches of northern Senegal's Mbour region.

After the Bissau-Guinean army offensive in early 2006, Sadio fled from the southern Casamance border with Guinea-Bissau (*Front Sud*) to the northern border with the Gambia (*Front Nord*). Sadio established his forces along the Gambian border and prepared for the coming fight with the Senegalese Armed Forces (SAF). In response, the SAF, having withdrawn from the Casamance after the 2004 cease-fire, in August 2006 reasserted its authority in the region by resuming old fortified positions and attacking Sadio's forces with armored vehicles and a few Mi-35 Hind helicopter gunships acquired earlier in 2006 from Ukraine.

The resumption of armed violence diminished hopes that peace and prosperity would return to the Casamance anytime soon. Attika—meaning

FIGURE **3.1.** Map of Senegal.

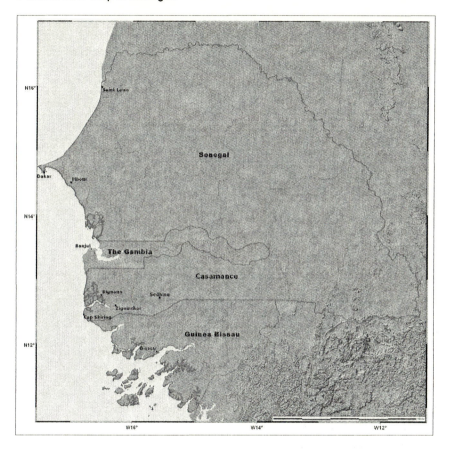

Source. Map created by Professor Peter L. Guth, Oceanography Department, U.S. Naval Academy from public domain data.

"warrior" in the Jola language—resumed its small-unit ambushes on Senegalese government vehicles, its deployment of land mines in roads and fields, and its highway robbery (literally) of travelers on Casamançais roads in search of cell phones, cash, and other petty resources. Nevertheless, the violence since 2000 has never been as bad as it was in the mid-1990s, at the conflict's peak. But while cease-fires are encouraging first steps, they are not lasting peace deals. Nobody labels the current banditry and low-intensity violence as peace, but it seems relatively tame next to the ferocity of the recent civil conflicts in Liberia and Sierra Leone. Consequently, the local press and diplomats in Dakar characterized the Casamance situation from 2005 to 2007 as "no peace, no war."

THE BITTER ROOT OF GRIEVANCE

On my first trip to Ziguinchor in 2005, the first explanation I heard for the cause of the conflict matched Lloyd's first driver. The Casamançais believed they had been neglected by the *nordistes* (northerners) in power in Dakar. More than that, they believed the *nordistes* had exploited them and their land in much the same way the European colonizers did. The Casamançais, and more specifically the Jola[316] ethnic group at the heart of the separatist movement found primarily in the Basse Casamance (the "Lower Casamance" that abuts the Atlantic Ocean), resented the flood of *nordistes* who claimed the majority of the jobs in the tourism sector that sprouted up on the Casamance's beautiful white beaches in the 1960s. Numerous local sources told me that the *nordistes* (primarily Wolof, ethnically) got most of the jobs because traditionally they were traders and herdsmen, accustomed to business.[317] The Jola, on the other hand, traditionally were agriculturalists who just wanted to be left alone to farm their land. However, once the Jola saw the *nordistes* making money in the tourism sector, they wanted in on the action, too, and resented their portrayal as backward farmers wholly unqualified for the new business climate.[318]

   Therefore, the Casamançais blamed this postcolonial history, with Dakar as the *nordistes*' capital, for the Casamance's poverty and broken infrastructure. They believed their potholed roads and mine-laced fields existed because Dakar just did not care. In broader terms, they believed Dakar had neglected the Casamance when doling out what Robert Rotberg calls the state's "political goods": roads, health care, education, economic improvement, security, and so on.[319] The MFDC made these accusations in spite of the posts held by Casamançais in the

Senegalese government—military and civilian. For example, President Wade in 2000 named Youba Sambou, a Jola from the Casamance, to be his armed forces minister.[320] In spite of this appointment and others like it, Casamançais politicians in Dakar were not able to bring home the proverbial bacon in terms of investment and development for the Casamance. Many Casamançais remained skeptical that the *nordiste* masters of the Senegalese state would ever allow Casamançais politicians to do so anyway.

In the seminal event that led to the establishment of the modern MFDC, Léopold Sédar Senghor allegedly struck a bargain with the leading Casamançais politician, named Emile Badiane, to garner support for Senghor's presidential bid against Lamine Gueye during Senegal's first presidential campaign in 1960. Gueye had secured Dakar and most of the *urbaniste* vote. Senghor, a Serer—considered a cousin ethnic group to the Jola (*le cousinage*)—needed the vote of the Casamançais and the rest of Senegal's rural political base. Therefore, Badiane and Senghor allegedly struck a deal that the Casamance would enter the postindependence era as part of Senegal as long as Senegal granted the Casamance its independence after twenty years.[321] The bargain is only alleged because no record of it exists. Michael Lambert asserts, "Some of my informants claim that Badiane, who died in 1972, was murdered by officials of the Senegalese government and that the Senegalese government subsequently destroyed the accord signed by Badiane and Léopold Senghor."[322] Hope for the independence of the Casamance seems to have gone to the grave with Badiane.

Well before the separatists' march on Ziguinchor in 1982, Father Augustin Diamacoune Senghor (no relation to Senegal's first president), leader of the political wing of the MFDC from the start of the conflict until his death in 2007, began detailing the MFDC grievances against the Senegalese state. In 1980 he wrote President Léopold Senghor, who was about to turn over power to Abdou Diouf, that "Casamance intends to fly on its own" (La Casamance entend voler de ses propres ailes). Father Diamacoune openly questioned the inclusion of the Casamance within the Senegalese nation-state, insisting that the "companionship" with Senegal had brought only "negative results" economically, politically, socially, culturally, and morally. Diamacoune went so far as to outline a plan for Casamançais "autonomy" and eventual "international sovereignty." In April 1981, he wrote another letter, proclaiming his "Casamance-ness" (ma *Casamancité*) and asserting that "the colonial power had not made of [him] a Senegalese" (le colon n'a pas fait de moi un sénégalais).[323]

THE BITTER ROOT OF IDENTITY: A COLONIAL LEGACY

As illustrated by Father Diamacoune's assertion of his *Casamancité*, essential to the legitimacy of the perceived grievance in Lloyd's model is the formation of a separate identity by the aggrieved group. The colonial legacy rendered this task much simpler for the Jola, who have essentially wrapped themselves in the larger Casamançais identity to legitimize their separatist actions while avoiding charges of "tribalism."[324] The modern nation-state of the Gambia provided the geographical separation for the Casamançais to claim a separate and unique identity from Senegal. To get to the bottom of that root, though, we have to go back to the nineteenth century and the efforts of Europeans to avoid going to war with one another over their friendly colonial competition in Africa.

Following the Berlin Conference—a series of meetings in 1884 and 1885 hosted by the German Chancellor Otto von Bismarck—European nations jockeyed for position in Africa, hoping to benefit from the solidification of vague notions of each nation's area of colonial interest. In the end, Europeans emerged from the process with the modern political map of Africa. Britain's Lord Salisbury remarked, "We have been engaged in drawing lines upon maps where no white man's feet have ever trod; we have been giving away mountains and rivers and lakes to each other, only hindered by the small impediment that we never knew exactly where the mountains and rivers and lakes were."[325] Though traders from several European nations traded freely on the Gambia River for centuries, the British dominated the trade after establishing a trading post initially at Fort James and later at the modern city of Banjul. Before the Berlin Conference, the French, who had established forts north and east of the Gambia River, wanted to unite their colonies in West Africa. Therefore, they attempted to negotiate with the British and Portuguese to consolidate their colonial holdings. Specifically, the French offered to trade Cote d'Ivoire and Gabon to the British in return for the Gambia.[326]

However, the British had already decided that trade on the Gambia was too lucrative to give up, so the Gambia became a British colony. The British designated the Barrakunda Falls as the interior limit of their colony on the Gambia River because beyond the falls, there were "a number of impediments to navigation and...the commercial prospects in those regions were disappointing."[327] Consequently, it was up to the Anglo-French Boundary Commission of 1889 to translate the lines on the Berlin Conference map into real boundaries on the ground between the Gambia and Senegal—leaving the Casamance separate and distinct from northern Senegal.

DESERTIFICATION AND IDENTITY

Desertification in West Africa added to this geographical—and eventually cultural—distinctness. While the rest of Senegal's arid landscape bordered the Sahara Desert, the climate of the Casamance provided apt amounts of rainfall to keep the region green and lush. As one Senegalese military officer told me during my first trip to the Casamance, "Tout pousse ici...même les pierres" (Everything grows here...even the rocks). Though the rest of the country's rainy season normally lasts two to three months, that of the Casamance lasts half the year, making it the rice basket of Senegal. While the countryside in northern Senegal is brown, mixed with splotches of gray brush and crooked, gray baobob trees, the Casamançais countryside consists of tall grass, palm trees, and deep, dark, green forest. Father Diamacoune effectively contrasted the imagery of the Senegalese baobab as compared to the Casamance palm tree:

> These strong images associate the destruction of the environment and Senegal's open spaces with inequality and violence as opposed to

> the "harmonious symbiosis of the land's rich hummus and the fertile
> wind of the open sea"; to the forest, the solidity, and the rituals and
> initiations of sacred forest shrines through which "the Senegalese
> did not pass"; and to Casamance democracy and egalitarianism.[328]

This symbolism demonstrates only one way in which Father Diamacoune and the MFDC used the Casamance's geographical and natural
distinctness to construct a cultural distinctness—and thus, a Casamançais identity. Father Diamacoune asserted that Casamançais culture was
so distinct from that of the rest of Senegal that it rendered assimilation
into the Senegalese nation-state impossible—similar to the argument
made by Canada's Quebecois, Ireland's Catholics, Spain's Basques, and
other separatist groups throughout history. However, Senegalese scholar
Hamadou Tidiane Sy shows that Casamançais identity came about as
a result of the conflict and not the other way around. He also identifies
three levels of "social or group identity" in the Casamance:

> Because the MFDC claims to fight for the independence of Casa-
> mance, the first level falls along regional and territorial lines,
> encompassing the region of Casamance. Because it recruits mainly,
> though not exclusively, within the Jola community, the second
> level must be defined along ethnic lines, encompassing the Jola
> ethnicity. Third, we must consider Senegalese civic identity as the
> "other," or enemy group, being fought by the MFDC.[329]

Of course, defining identity vis-à-vis the other is essential to building
the myth or the logic—depending on your view—of the separatist argument: "We are different from them; we have a right to govern ourselves;
they have no right to rule us."[330]

As a personal illustration of Sy's three levels of Casamançais identity,
Cecile worked for my family as a cook and nanny while we lived in Dakar.
A middle-aged Jola woman, educated by Catholic nuns in a Canadian
missionary church in the Casamance, married to a Senegalese military
officer and later divorced, she had lived and worked in Dakar since the
conflict began in the Casamance. I took Cecile and my family with me on
my second trip to the Casamance (December 2005—one year since the

signing of the cease-fire). We drove our family SUV and planned to drop off Cecile for a visit to her home village just west of Bignona. Before the trip, I asked Cecile about the multiple layers of her identity. The following is my recollection of what she told me: "When I am in my home village, I am Jola. When I am working in Dakar, I am Casamançais. But when the Senegalese national soccer team is playing in the Africa Cup, I am Senegalese." Cecile's response tells me that her identity is situational and flexible. It also tells me that at least a part of it is constructed.

HISTORICAL ARGUMENTS

Historical arguments have played a key role in the formation of Casamançais identity, as the conflict's antagonists have emphasized different interpretations of the region's history. Since traditional African storytellers, or *griots* as they are known in West Africa, passed much of Africa's history from generation to generation—especially in those areas absent strong states or not influenced by Islam—little history was actually written down until Europeans arrived. Consequently, in Senegal, where one stood on the Casamance issue depended largely on whether one emphasized the French or the Portuguese colonial history.

Since Europeans considered the Casamance a part of Portuguese Guinea until 1888, the MFDC leaned more toward Guinea-Bissau than Senegal and rested its historical argument that the Casamance was separate and distinct from the rest of Senegal on the Portuguese colonial history. Father Diamacoune appealed to this history in a letter to a local newspaper in 1990:

> According to Honore Pereira Barreto, Governor of Portuguese Guinea, based on documents he consulted, the Casamance was the first river and the first country the Portuguese crossed on the western side of black Africa…The Portuguese administration was established at Ziguinchor in 1645, fourteen years before the foundation of Saint-Louis in Senegal by the French in 1659. It remained there until 1888, or 243 years of Portuguese presence. This is not forgotten.[331]

Father Diamacoune asserted that the Casamance became a part of Senegal only when the French colonial administration decided to include it, shortly before independence in 1960.

Previous to that, French settlers in the Casamance, building on the Portuguese legacy and the geographical separation of Senegal by the Gambia, acted autonomously with the *métropole*, refusing to deal with Paris via the colonial administration in Dakar. That no infrastructure connected the Casamance to northern Senegal made it easier to do so. Consequently, French colonial merchants traded directly with Paris from the Casamance instead of sending their goods and produce through the Port of Dakar.[332] Claude Michel, the colonial administrator superior of the Casamance, wrote in his annual report of 1944 that there was a growing sentiment among French settlers and indigenous Casamançais that "particularism [and] regionalism dominate in all locations. One is Casamançais and French first, Senegalese with misgiving."[333]

However, once the Trans-Gambian Highway was built in the late 1950s and Senegal gained its independence in 1960, the process Senegalese historian Mamadou Diouf calls the "islamo-wolofization"[334] of Senegal began, as *nordistes* asserted state authority in the Senegalese periphery—including the Casamance.[335] Suddenly, the Jola faced a new "colonizer": Senegal. Many Jola began to regard the Jola tradition, or *mujolooayi*, as the opposite of *mululumayi*, or "the tradition of the whites," referring not only to the European way of doing things but to the Dakarois, or *nordiste*, way of doing things as well.

> In effect, when the Jola say mululumayi, they speak of the nordiste way of working, the nordiste way of dress, the nordiste way of commanding the rest of the country, the nordiste manner of construction: of everything defining the white man's way of doing things.[336]

Equating the *nordistes* with the French provided about as visceral an argument as one could provide in postcolonial Senegal for Casamançais separatism.

The Portuguese history better facilitated the MFDC's separatist argument. As Diouf points out,

> the reference to Portuguese history carves out a territorial space. The map of the Jola Republic, drawn by Attika…included the Lower Casamance and a part of the Republic of Guinea-Bissau, namely the province of San Domingo, in the Cacheu region.[337]

While several modern writers have correctly drawn attention to the factionalization and the diversity of the Jola, the "Jola nation" extended from the Casamance into Guinea-Bissau and the Gambia.

REGIONAL SOLUTIONS FOR REGIONAL PROBLEMS

The irredentism of the Jola nation (the fact that it spans the borders of three modern nation-states) explains why so many of those working on peace in the Casamance appeal to the Senegalese government to treat the "Casamance question" not as an internal affair but as a regional one. For MFDC spokesman Famar Goudiaby, finding a regional solution with the help of international mediators is the only way forward, if only because of the 60,000 displaced persons in the Gambia and Guinea-Bissau.

> The Casamance conflict has overflowed Senegal's borders and it is imperative to involve foreign countries in settling it…to solve the problem internally will never be a solution…Guinea-Bissau and Gambia cannot be circumvented in the peace process…a tri-government solution is the only solution.[338]

The governments of Guinea-Bissau and the Gambia have not always helped matters a great deal. In the early years of the conflict, the *maquis* fought Senegalese forces with bows and arrows. As the SAF was a modern, conventional force trained and equipped mostly by the French, the fight was fairly asymmetric. However, as small arms and munitions began to flow out of the former Soviet Union as well as the West, to a lesser extent, at the end of the Cold War, the MFDC acquired deadlier weapons, including Kalashnikov rifles, mortars, rocket-propelled grenades, and land mines. As one businessman in Guinea-Bissau told me, "It was easier [for the Cold War powers] to sell or give away the arms than it was to properly dispose of them."[339] In 1998 the president of Guinea-Bissau, João Bernardo "Nino" Vieira,[340] accused his former armed forces chief of staff, Brigadier General Ansumane Mané, of trafficking illicit arms to the Casamance to support the MFDC. In response, Mané, a Mandingo, led the majority of Guinea-Bissau's armed forces in a coup d'état to remove Vieira from power. The government of Senegal, understandably upset that members of the military junta in Guinea-Bissau had been supporting the MFDC, ordered its armed forces to invade Guinea-Bissau to restore Vieira to power. The Senegalese failed in this mission and eventually withdrew from Guinea-Bissau after an Economic Community of West African States (ECOWAS)-brokered cease-fire. Meanwhile, Attika continued to traffic arms, precious woods, and cannabis from sanctuaries just across the border in Guinea-Bissau (*Front Sud*)—until the 2006 combined Bissau-Guinean and Senegalese offensive that drove Salif Sadio north toward the Gambian border (*Front Nord*). As long as Sadio's forces continue to find sanctuary in the Gambia and Guinea-Bissau, they will have the tactical flexibility to continue the struggle. Lord Salisbury's lines on a map will not stop them.

Senegal's relationship with the Gambia has always been problematic. As another cruel joke of the colonial legacy, the president of the Gambia since 1994, Yayah Jammeh, is Jola, which inherently complicates Senegalese perceptions of Gambian words and actions. President Jammeh has consistently denied that he knowingly provides sanctuary to the MFDC, but Gambian efforts to effectively control its southern border have often appeared meager and ineffective. Furthermore, Gambian arrests of MFDC members have proven counterproductive to the peace process, as when Gambian authorities arrested Magne Dieme, one of the two leaders of Attika willing to break with Sadio and negotiate with the government of Senegal, in 2006. Jammeh used the arrest to proclaim that he was helping Senegal by arresting one of its rebels, when in fact he was only removing one of the few rebel leaders willing to negotiate.

President Jammeh has had an interest in perpetuating the conflict to keep Dakar off balance. Jammeh understandably feels insecure, surrounded as he is by the larger nation of Senegal. He may also feel threatened by Senegal's status as a stable, multiparty democracy since independence and its being one of the few African countries that has never had a coup d'état. Jammeh, a Gambian army officer in 1994, seized power in a coup. He now seems to believe that if he truly works with the Senegalese government to facilitate peace and that if he builds a bridge across the Gambia River to facilitate traffic on the Trans-Gambia Highway (in 2005–2006, truckers waited for days to have the opportunity to board the one ferry in operation), he will lose the leverage he ostensibly has to force Senegal to pay attention to him. Jammeh realizes how much Dakar stands to benefit from peace and development in the Casamance. Therefore, it seems he wants to slow Senegal's development to keep his cousins to the north in the gutter with him.

THE BITTER ROOT OF POWER: COMMITMENT AND RESOURCES

The third bitter root combines the practical with the mystical in the case of the MFDC. A Jola once told me that to understand the Casamance conflict, one must go into the Bois Sacré—the Jola's Sacred Forest. The MFDC's power has been derived in part from mystical forest ceremonies in which participants take oaths to their ancestors to win independence for the Casamance or to die trying. The Bois Sacré has other purposes as well—meting out discipline within Jola society, for example—but when it comes to explaining why a Jola picks up the gun, one must appreciate the importance of the oaths. The Sacred Forest is a special place in Jola culture—something well known by all sides in the conflict. It is the place from which the MFDC demonstrators in Ziguinchor in 1982 commenced their march after taking their oaths.[341]

But few in the peace process have given the Sacred Forest the importance it deserves.[342] That may be in part due to the difficulty of entering the Sacred Forest society. One must first be initiated. And the initiations take place only once in a generation—infrequently, at best—so there have been talented Casamançais politicians who spoke with authority in the eyes of Westerners but had no authority with traditional Jola because they had little or no experience in the Sacred Forest.[343] Consequently, the Senegalese government and Western mediators, because they failed to acknowledge the importance of the Sacred Forest to the Jola, have approached the wrong members of the MFDC in the peace process. Empowerment in modern, Western political structures is not the same as empowerment in a traditional culture like that of the Jola.

Despite the power of these traditional beliefs, Casamançais commitment began to wane following the 2004 cease-fire. Casamançais villagers simply grew tired of the fighting, the land mines, and MFDC demands for food, money, and protection. Besides, displaced villagers began to return to their lands; hope for a lasting peace was on the rise. I wondered, could the *maquis* not somehow renegotiate their oaths to the ancestors so that they could accept some sort of autonomy short of full independence? However, the *maquis* were more interested in a different question: as Casamançais commitment began to wane, could the MFDC resource base make up for the overall loss in power so that the *maquis* could continue the struggle? What seems counterintuitive about the MFDC's resource base is that its weakness has been its strength—at least in terms of the conflict's longevity. The MFDC's relatively weak resource base has perpetuated the conflict because it has perpetuated the notion that the conflict is "not that bad."[344] With no "blood diamonds" or oil wealth to pilfer, what was there to fight over? Not much. Consequently, the MFDC trafficked marijuana, valuable hardwoods, and some cashews—in addition to weapons—to bankroll its activities. As one reporter noted, "While the profits are modest, in an impoverished region, they are better than nothing."[345]

However, growing Latin American drug cartel involvement—especially in Guinea-Bissau in 2006 and 2007, but increasingly in Senegal in 2008—signals trouble on the horizon. A November 2008 report by the UN Office on Drugs and Crime detailed the scope of the threat:

> In the last three or four years, it has become clear that the special vulnerability of Africa is being exploited. Cocaine produced in South America is increasingly taking a detour on its way to

growing markets in Europe. The area most affected is West Africa, a poor region recovering from many prolonged civil conflicts... The relationship between diamond smuggling and the civil wars in Sierra Leone and Liberia has been well documented, but, at their peak, profits accruing from this activity amounted to some tens of millions of dollars per year. The potential destabilizing influence of the cocaine traffic, where the value of a single consignment can exceed that sum, is very real. The profits generated by this trade are larger than the entire security budgets of some of the smaller West African countries. But it is likely the traffickers are making use of the region for a far smaller sum, paid out in kind, rather than cash. This has generated local cocaine markets, which pose hazards of their own.[346]

As if the threat from the drug trade's inherent corruption was not enough, one can now add cocaine itself as a potential catalyst to the Casamance conflict. If Casamançais begin to consider drug trafficking a viable survival strategy against the perceived discrimination and negligence of the Casamance by the Senegalese state, one can anticipate more violence at a higher level of intensity in the future.

THE AID ENABLER?

In another bitter irony, bilateral and multilateral donors, as well as the Senegalese government, may have perpetuated the conflict by their generosity. Payments from the government of Senegal to MFDC rebels who participated in the peace process, in addition to bilateral and multilateral aid, gave the *maquis* a reason to keep fighting. The Senegalese government began providing payments through Major General Abdoulaye Fall, the chief of staff of the Gendarmerie Nationale, in 2005, following the December 2004 cease-fire. General Fall ostensibly handed out the payments to encourage the *maquis* to quit the forest and participate in the peace process, but accusations of corruption led to questions about the payments' effectiveness.

Though equally well-intentioned, Western donors may have "contributed" to the conflict as well. Numerous Western governments, international organizations, and NGOs provided jobs and assistance to Casamançais working on peace and development in the Casamance. They called for meetings, dinners, and receptions with local civic organizations and the MFDC itself. Martin Evans, a geographer at the University of Leicester, admits that bilateral and multilateral aid raises the question of what everyone in the region will do once the conflict is over: "Since 2000, there's been a lot of multilateral and bilateral money coming in to support the peace process, return of the displaced and reconstruction...it can be a bit of a gravy train for everyone."[347]

The U.S. military must also be careful not to aggravate the situation with its security assistance. During my tour in Dakar, representatives from U.S. European Command (EUCOM)[348] approached the Senegalese government with counterterrorism security assistance, with the caveat that the assistance could not be used against the MFDC but only against "transnational terrorist groups." Senegalese officers responded that the MFDC operated in at least three different West African countries and in Europe as well. It also planted land mines on roads intended for civilian use (as in the small bus that struck a land mine in northern Guinea-Bissau in 2006, killing twelve civilians)[349] and terrorized Casamançais villagers for money, supplies, and information. How could U.S. officers deny that these actions were those of "a transnational terrorist group"? the Senegalese officers asked.

In response, EUCOM officers stressed that the MFDC was not a jihadist transnational terrorist group in the mold of Al-Qaeda in the Islamic Maghreb (AQIM), at the time known as the Salafist Group for Preaching and Combat (GSPC). EUCOM saw little success in its efforts to curtail the radicalization and expansion of this Algerian-grown group through its Trans-Sahara Counterterrorism Partnership.[350] If anything, the United States' focus on AQIM may have added to its recruitment, prestige, and power.[351] Therefore, EUCOM made this distinction to the SAF regarding operations against the MFDC, an important distinction to Senegalese perceptions of the peace process—in the Casamance and in Dakar—that the United States was an honest, independent broker.

On the other hand, Western assistance has played an important role in the peace process. From 1999 to 2003, the U.S. government spent 9.1 million U.S. dollars on programs related to "peace and security" and "conflict resolution" in the Casamance.[352] The Canadian and American governments contributed hundreds of thousands of dollars to the region for humanitarian demining. Some of this money may have contributed to the aid "gravy train," but some of it helped bring about the December 2004 cease-fire as well. Nearly every person I talked to in the Casamance told me that the U.S. embassy—particularly U.S. Ambassador Richard Roth—played an indispensable role in bringing the MFDC and the Senegalese government together to sign the cease-fire. And, most added that this was a role only the Americans could have played because of their influence and their lack of colonial baggage in the region.

THE BITTER ROOT OF FACTIONALIZATION: DESIGNATING A VALID SPOKESPERSON

Lloyd's final requirement for conflict has always been the most troublesome for the MFDC: designating a valid spokesperson. To conduct peace negotiations, the government of Senegal and Western mediators had to know whom to talk to in the first place. The *maquis* were so factionalized that it was often difficult to tell who legitimately spoke for the MFDC. Even before the death of Father Diamacoune in 2007, many working in the peace process—including myself—often posed this question. It seldom appeared as if the MFDC's political wing under Father Diamacoune and its military wing—Attika, and particularly Salif Sadio—agreed on a way forward. Once Father Diamacoune passed away, Nkrumah Sané, the MFDC representative in Paris, and Ansoumana Badji, formerly the MFDC representative in Lisbon, vied for the title of being the new MFDC spokesperson. In reality, however, they had little power to effect change in the Casamance because Sadio had all the guns, and neither of them had the venerable status of the late Father Diamacoune.

Sadio seems to have assumed the mantle as the MFDC's valid spokesperson, as he remains the only one with the power to give or refuse battle. Whether the MFDC agrees on his role as a valid spokesperson or not, the local press covers every word he utters and every move he makes because he is the "last of the holdouts" and the one causing all the trouble for Senegalese security forces. Among many Casamançais and especially among the Jola, his recalcitrance only further burnishes his bona fides as a *maquis* and a Jola of great faithfulness to the original founders of the movement.

Father Diamacoune's passing in 2007 marked a real transition and a blending of roles between the armed wing and the political wing of the MFDC, which was bitterly divided at the time. As Badji and Sané argued on the radio and in the newspapers over who was running the political wing, Sadio vanquished the other factions of Attika and consolidated his power over the military wing of the MFDC.[353] Since the political wing seemed so ineffective—not least because Sané was speaking from Paris—and, as mentioned, because Sadio alone had the power to give or refuse battle, Sadio embodied not just military but de facto political power for the *maquis*. Will this blending of the MFDC's political and military wings in the person of Salif Sadio lead to better coherence and less factionalization of the *maquis* overall? Perhaps in the short term. But in the long term, concentration of more power in one man can only endanger the longevity of the institution of the MFDC overall.

CONCLUSION

According to Lloyd, most conflicts end with one side winning unilaterally, and the winner is usually the state. This will be the most likely outcome in the Casamance. Of course, we have seen numerous conflicts in Africa where the rebels join or become the state, as in the case of Charles Taylor's National Patriotic Liberation Front (NPFL) in Liberia or John Garang's Sudanese People's Liberation Movement (SPLM) in Sudan. The other potential outcome is what Lloyd calls a "mutually hurting stalemate." This seems to be the current situation in early 2009—that

I called earlier "no peace, no war." While direct confrontation between the MFDC and the SAF seems to have ended since 2007, one still reads of occasional (monthly) violence initiated by the *maquis* that blurs the line between banditry and political violence. Furthermore, Salif Sadio and his fighters are still in the forest. Until Sadio is killed, captured, or brought to the negotiating table, a final peace agreement is unlikely.

In summary, I have asserted that the *maquis* of the Casamance have wrapped their predominantly Jola identity in the larger Casamançais identity to continue fighting for independence from Senegal until the cost of doing so outweighs the benefit. The drivers—the bitter roots—of the Casamance conflict result in large part from the legacy of colonialism in the region. Specifically, they result from Casamançais grievances over discrimination in the postindependence tourist sector and a sense of neglect over the "political goods" of the Senegalese state, such as decent roads to help realize the strong agricultural potential of the region. Villagers' dwindling commitment to the goals of the MFDC and their fatigue with the conflict in general, in combination with the factionalization of the MFDC as a result of arguments about who should speak as the legitimate spokesperson for the movement, may be signs that the conflict is winding down. However, the historical weakness of the MFDC's resource base, combined with the potential for South American drug involvement and Salif Sadio's embodiment of the political and military wings of the MFDC, points to the potential for West Africa's longest-running civil conflict to continue well into the future.

Chapter 4

Resource Allocation and Xenophobic Violence in South Africa

Anthony Turton

Introduction

South Africa is a complex country with a rich and convoluted past. Running through the modern history of the country like a fine golden thread is the central issue of access to resources. Two sides of this coin are evident in this chapter: resource capture and reallocation of resources, which form a rich backdrop to an analysis of the transition from oligarchic rule to democracy, such as happened in 1994. This transition was a dramatic watershed event in modern South African history, so it provides an appropriate backdrop against which the emotive issue of violence can be dealt with, more specifically, as it might relate to a shift from resource capture to resource reallocation. A hypothesis is presented for future validation—when the expectations of society for resource allocation exceed the capacity (or willingness) of the government to deliver, then mass violence could result.[354] Water is the single most important resource that constrains the future economic development potential of the country, so it is an appropriate subject of study. This will be done in the context of what the author has called the Uhuru Decade, which is that period of history following a liberation struggle, during which the victorious liberation movement inherits infrastructure and institutions to do with as it chooses.[355] This Uhuru Decade thus provides a neatly defined window of time in which transition can be studied with a view to determining if resource reallocation or persistent resource capture will be the foundation for future policy.[356]

RESOURCE CAPTURE AS AN ELEMENT OF SOUTH AFRICAN HISTORY

South Africa was born as a sovereign entity out of the ashes of the Second Anglo-Boer War (1899–1902).[357] Prior to this watershed event, the region known as South Africa consisted of four separate entities, two being British colonies (Cape and Natal) and two being independent sovereign states known as the Orange Free State and the Transvaal Republic. The latter two states were Westphalian in origin and were the home to Boer people, descendents of the original Dutch settlers to the Cape of Good Hope in 1652, who migrated away from British control during the Great Trek (1835–1840), after the hinterland had been depopulated by a cataclysmic event known as the Mfecane.[358] The Mfecane was an event of extreme violence, which today would be described as ethnic cleansing, when the Zulu nation rose up between 1816 and 1828 and swept across the region decimating local tribes and capturing cattle. The word "mfecane" means "to be crushed in total war." Welsh describes the Mfecane as leaving

> a generation of homeless refugees…doomed to wander about South Africa seeking temporary refuge and food, but being driven to cannibalism and starvation. The passage of such armies led to

widespread devastation. European writers spoke of travelling for days through a deserted countryside past the scattered bones of the dead. Into the geographical void left by the killing and dispossession moved the...[Trekboers], the pioneers of what became the Afrikaner [Boer] republics.[359]

The Second Anglo-Boer War was a bitter affair, driven in essence by a pompous warmonger named Sir Alfred Milner, who had arrived in the Cape Colony in May 1897 as the new British high commissioner. He had a powerful personality and very specific views about empire, so his appointment by the British government was set to make a major impact in the region called South Africa, even though it was not yet a country. Meredith calls Milner an "imperial zealot," who described himself thus:

> I am a British Nationalist. If I am also an Imperialist, it is because the destiny of the English race, owing to its insular position and long supremacy at sea, has been to strike fresh roots in distant parts of the world. My patriotism knows no geographical but only racial limits. I am an Imperialist and not a Little Englander, because I am a British Race Patriot.[360]

In this role as "a civilian soldier of Empire," "Milner possessed a formidable intellect but a narrow mindset," notes Meredith.[361] This appointment thus confirmed the prevailing British belief that the ultimate objective of their policy in southern Africa should be to steer the Transvaal into the imperial sphere before it became powerful enough to determine its own destiny in the region, because that would not favor the British.[362] Milner can be regarded as being a cultural Darwinist, believing as he did that natural selection had favored the British race over time. He despised the Boers, whom he considered to be a degenerate race of Europeans who had became backward because of their closeness with the indigenous African people (the word "Afrikaner" means "to be African").

The reason for their fear of a growing Transvaal hegemony was the discovery of gold in 1886, which the British coveted and decided they wanted to control at all costs. The actions that flowed from this

appointment, most notably the Second Anglo-Boer War, with the subsequent scorched-earth policy and British concentration camps as a key
component, should thus be interpreted against this historic background.
The scorched-earth policy, much like the earlier Mfecane, was a brutal and dehumanizing affair. It arose as official British policy when the
capital cities of the two Boer republics (Orange Free State and Transvaal) were captured in 1900, but the Boer people still refused to con-
cede defeat. A number of Boer generals led what became known as the
Bittereinder Commandos—the word "bittereinder" meaning "those who
chose to fight to the bitter end"—who started to fight an unconventional
war based on guerrilla tactics in the face of asymmetrical power and a
rigid British military doctrine. This is also the origin of the word "com-
mando," which is now part of the global military lexicon.

This new style of warfare so infuriated the British high command that
they decided to burn the farmsteads and crops of the Boer people in a
bid to destroy their will to fight. This has been eloquently described by
soldiers responsible for the burning.[363] The scorched-earth policy drove
the Boer women and children off the land and into British concentration camps, which eventually killed more people (women, children, and
noncombatant black South Africans who were seen to have collaborated
with the Boer people) than deaths on the field of battle on both sides.
Welsh cites October 1901 as being the peak of deaths in the concentration camps, with a rate of 34 percent.[364] Meredith cites the death toll in
the camps as being mostly among women and children under the age
of sixteen, amounting to one-tenth of the total Boer population in the
Orange Free State and Transvaal at the time.[365] In separate camps for
black South Africans who were deemed to have collaborated with the
Boer people, the eventual prison population was 116,000, out of which
some 14,000 died of disease and malnourishment. Details of these concentration camps were made known to the British public by Emily Hobhouse,[366] a Quaker woman whose reports were initially refuted on the
grounds of her being "hysterical"[367] but were later found to be accurate by the Fawcett Commission of Enquiry[368] and therefore taken into

consideration by the British government as it changed its policy to a more humane one.[369]

The importance of the Second Anglo-Boer War in the context of this chapter is largely twofold. Firstly, it was, in essence, about resource capture. In this case, it was mineral wealth, in particular gold and diamonds, that was the main driver,[370] but the end result was the capture of a resource that ultimately gave power and privilege to those who controlled it.[371] Resource capture thus became a key element in the power politics of a future South Africa, which has endured to this day.[372] Secondly, it was an extreme event, which magnified the impact of an earlier period of even greater violence—the Mfecane—which in effect hardwired the experiences of social trauma into the South African collective conscience. The author is of the opinion that the experiences of having been the victims of violence are so deeply etched into the national psyche of the South African people that this plays a role in contemporary political dynamics, and thus national security, to this day. More importantly, however, is the shocking reality of what happens when the victim later becomes perpetrator, such as occurred during apartheid and is starting to occur again under majority rule.[373] Central to this is the notion of human rights and environmental justice, which is an extremely sensitive issue today, as evidenced by the banning of the author for raising it in a recent scientific conference.[374]

Water as a Critical Natural Resource

Control over water is as important as control over mineral resources. South Africa provides a classic case of resource capture in the water sector, where the Afrikaner (Boer) minority stayed in power during the period of apartheid rule by controlling water allocation as a source of political patronage,[375] while it also collaborated with, rather than regulated, the gold mining industry[376] in order to sustain minority rule in the face of protracted economic sanctions and isolation as a pariah state.[377] In water-constrained political economies, a simple equation can be used to demonstrate how the best potential can be unlocked from the total available stock of water. This is expressed as follows:

$$V \times F = Y$$

where V is the stock value of water, expressed as the volume of the total national water resource that is available at a high assurance of supply level at any one time, F is the multiplier value of water as a flux, and Y is the total water supply needed at a high assurance of supply to sustain the desired economic development potential of the country at any moment in time.

From this simple equation, it becomes evident that if V is a constant (because almost all readily available sources have been developed, such as in South Africa),[378] with a value of, say, 10 × 109m3yr1 (10 billion cubic meters of water per year), and if F is 1, then the value of Y would be 10 × 109m3yr1. Stated simplistically, then, the total economic development potential of the country dependent on water (Y) would be constrained by the volume of water available at a high assurance of supply (V). If, however, the F value is ≤ 1, say, for argument purposes, 1.2, then a different outcome is possible, because the value of Y then becomes 12 × 109m3yr1. Now, assuming the same logic applies, if the F value becomes 2, meaning that the multiplier value of water as a flux allows the total national stock to be used twice, then the value of Y becomes 20 × 109m3yr1, or double that of V. Seen in this light, it is the flux value of water (F) that becomes the determining factor of the economic development potential of a water-constrained country, and not the actual resource available at a high assurance of supply (V).

The South African case is presented in figure 4.1 below. This shows that the total national water resource available at a high assurance of supply based on current technology is 38 × 109m3yr1. Even on a low water-use estimate, the future demand for water by 2035 will be 65 × 109m3yr1, which leaves a deficit of 27 × 109m3yr1.[379]

If one applies the equation V × F = Y, assuming that the total surface resources available within the constraints of existing technology are 38 × 109m3yr1 and the CSIR projections are valid, then the future

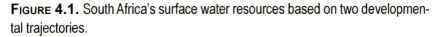

FIGURE **4.1.** South Africa's surface water resources based on two developmental trajectories.

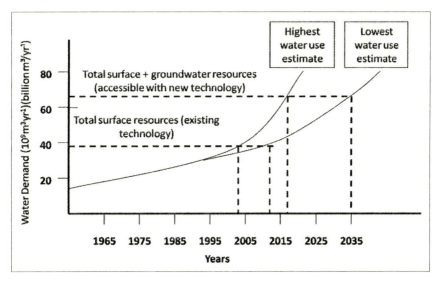

Source. Redrawn from CSIR, "Is the Glass Half Full or Half Empty?" *Science Scope* 3, no. 1 (2008): 19.

Note. The availability of South Africa's surface water resources is based on two developmental trajectories, showing that even with lower water use, the country has reached the limit of its surface water resource.

multiplier value of water as a flux will have to be 1.7105 in order to meet the needs for the economic development of the country, either by 2020, if a high developmental trajectory is to be followed, or by 2035, if a low developmental trajectory is to be followed.[380] This can be expressed as follows:

$$(38 \times 109m3yr1) \times 1.7105 = (64.999 \times 109m3yr1)$$

In reality, however, this is not happening, because the rapid changes caused by the transition to democracy have caused a massive deterioration of the water quality in most major reservoirs,[381] often the result of lost technical capacity caused by aggressive affirmative action policies and the resultant migration of young engineers out of the country.[382] While no accurate study has been done to quantify the impact that this deteriorating water quality has on the value of water as a flux, it has been roughly estimated by the author as being in the order of $F = 0.8$, which means that the value of $Y = (30.4 \times 109m3yr1)$. The source of this deterioration can be generically broken down into three main categories of pollution: radiological, chemical, and biological.

Radiological contamination is arising from uncontrolled decant of acid mine drainage from abandoned gold mines.[383] This is happening because the geology of the gold-ore bodies is also closely associated with uranium and other heavy metals, some of which are radioactive.[384] The city of Johannesburg has been built on top of a geological feature that now contains a number of hydrogeological compartments that are all filling up with contaminated water, some of which have already started to decant to the surface as mines close down,[385] but the rest of which are expected to reach decant levels in the next decade. This is a major crisis.

Chemical contamination has a number of sources, including endocrine-disrupting chemicals (EDCs);[386] acid mine drainage from coal mines, mostly manifesting as a heavy sulphate load;[387] and gold mines, also manifesting as a heavy sulphate load but including a range of heavy metals, some of which are radioactive.[388]

Biological contamination is most acutely being expressed as eutrophication, which is a result of dysfunctional sewage treatment plants that are allowing masses of nutrients to enter the major rivers and reservoirs of the country.[389] The effect of this is to reduce the value of the current stock of water (V), because it renders that water no longer fit for any purpose. Carol Paton reports that agricultural exports to the European Union currently valued at ZAR 28 billion (US$3.5 billion) are under threat because of high microcystin levels in the irrigation water, to name but one example.[390]

POSTDEMOCRATIC VIOLENCE

While the transition to democracy was remarkably peaceful under the circumstances of the prevailing armed liberation struggle that it ended,[391] there have been two subsequent periods of extreme violence in South Africa. The first was a series of riots that occurred in Phumelela and Merafong City in 2004, which resembled phases of the liberation struggle. In a research report commissioned by the treasury, the causes of this violence were found to be many and varied, but significantly, failure of service delivery, poor governance, lack of government capacity, corruption, and the dolomite water issue were highlighted as being specific drivers.[392] The second was a brief period of violence in 2008, the likes of which had rarely been seen, even during the darkest days of the armed struggle to liberate South Africa from apartheid. That violence occurred when frustrated citizens, most of whom were unemployed and seemed not to have benefited from the transition to democracy in 1994,[393] rose up spontaneously and murdered foreign nationals, often by setting victims alight and watching them burn in an orgy of pure barbarism.[394] While the uprising against foreign nationals was deplorable, it was the type of violence that was the most shocking, because in many cases, people were stabbed, bludgeoned, and then set alight in ways that almost resembled the type of nonindustrialized killing that occurred during the Rwandan genocide. It was the closeness between the victims and perpetrators, in a social sense, but also the way that the final killing was done that was the most shocking.

The significance of the xenophobic violence lies in four distinct factors: Firstly, it was totally unpredicted and thus, unexpected. It literally came out of nowhere and descended on various groups of angry people, who were instantly transformed into violent mobs with a bloodlust possibly only matched during the Mfecane. Secondly, it was driven, at least in part, by failed government policies, specifically, the reallocation of resources to the marginalized majority, who had not seen many improvements to their own lives since the end of apartheid rule, despite the great promise that democracy originally held for them.[395] Thirdly, part of this disillusionment has been driven by growing levels of corruption, where political elites have used their new position of authority to enrich themselves at the expense of those they were elected to represent.[396] Finally, it has been accompanied by extreme sensitivity by government officials, who simply refuse to hear anything that might suggest that their own policies might be a cause of the shameful events. This was made manifestly clear to the author when he tried to present the above argument to a scientific conference at the Council for Scientific and Industrial Research (CSIR), to which he had been invited as a keynote speaker in November 2008. The response was brutal: the author was banned from the premises and suspended a few days later, being ordered to leave with immediate effect, only to lose his job altogether as the event hit the media and the resultant public outcry was interpreted by the executive as bringing the CSIR into disrepute.

Postapartheid South Africa is thus not yet freed from the ghost of spontaneous violence, and it is quite probable that the collective experiences that are etched into the minds of the national psyche have not yet been effectively exorcised. This might account for the extreme sensitivities that arise, supported by the deep-seated denial that blocks any attempt to raise the issue in circles that are normally rational and erudite, suggesting that there is unpredictability about the whole business that deserves to be better understood.

RESOURCE REALLOCATION
OR CONTINUED RESOURCE CAPTURE?

The hypothesis developed by the author in his banned paper was that when the expectations of society for resource allocation exceed the capacity (or willingness) of the government to deliver, then spontaneous mass violence could result.[397] This still seems plausible to the author in the face of his banning and is therefore offered here to a more international community of scholars in the hope that it might be tested empirically.

When South Africa went through the democratic transition in 1994, the embattled and isolated country had been sustained largely by revenues generated from the gold mining industry. Those revenues are under threat because the gold industry is now mature and many mines have reached the end of their useful economic life. The legacy of gold mining is now being manifest as deteriorating human health amongst the poorest and most marginalized members of society, and the current government will have to deal with that complex reality one way or another. Underpinning the choices that the democratic government can make are two irreconcilable forces, and the way that these are dealt with will, in all probability, determine whether South Africa becomes a stable multiparty democracy or reverts back to the authoritarian rule of a one-party tyranny, such as happened in Zimbabwe. These two forces can best be understood as the need to reallocate resources, on the one hand, if political stability and social justice are to prevail, versus continued resource capture by the new political elite, thus perpetuating the deeply entrenched history of plunder for personal gain, now with a new, racial complexion. This has been described in considerable detail by the author elsewhere, and space precludes more analysis here.[398]

A golden opportunity exists in the area of mine water decant, which is associated with radioactivity and heavy metals. The significance is that the decant has already begun, and it can only accelerate as more mines close down and the mine void under Johannesburg, claimed by some miners to be as large as eight times the volume of Lake Kariba, fills up and starts to overflow in the various rivers that eventually become the drinking and industrial water sustaining the largest single economy in Africa. After all, the Gauteng province, in which Johannesburg is located, generates 10 percent of the economic output of the entire African continent, employing 25 percent of all South African citizens, while being 100 percent reliant on interbasin transfer of water.[399] More importantly, however, there is sufficient anecdotal evidence to suggest that human health is already being impacted,[400] with the most logical pathway of either the radionuclide or heavy metal load into infants being the cultural practice of geophagia, also known as pica, when pregnant women ingest clay found alongside streams to supplement minerals that their bodies need. It was the author's suggestion that this cultural practice might trigger potential violence in future if people start to believe that the government has been aware of hazardous waste in sediment and clay residue alongside contaminated streams and failed to do something to ensure human safety, which triggered his recent banning and subsequent suspension.[401] The issue is clearly one of extreme sensitivity that deserves to be researched in a balanced and robust way in the future. The author is unable to continue with this research as he has lost his academic home and is probably unemployable in any public research institution in South Africa by virtue of the government control of such appointments.

CONCLUSION

South Africa is a water-constrained country, and future economic development will be curtailed by this fundamental fact in the near future. Resource capture has been a fundamental component of the political dynamics of the country for more than a century, with very little empirical evidence that this is about to change. On face value, the reallocation of water resources has been accompanied by a catastrophic collapse in sewage treatment infrastructure, which now threatens to shrink the remaining national water resources needed to sustain economic activity to around 80 percent, but this is an arbitrarily defined number, used by the author merely for illustrative purposes. More alarming is the observed fact that resource capture in the mineral sector seems to be as deeply entrenched now as it ever was under apartheid. If the work by investigative journalists like Adam Weltz is to be believed, then there is some evidence to suggest that the author's banning and subsequent suspension and dismissal might have been related to the work he had been doing into the human health impacts of radioactive and heavy metal contamination found in decanting mine water.[402] In this regard, Weltz notes that "if a connection were found between gold mining and illness among the local populace, mining houses might be liable for massive damages.[403]

The hypothesis—when the expectations of society for resource allocation exceed the capacity (or willingness) of the government to deliver, then spontaneous mass violence could result—as proposed by the author in his banned paper, has yet to be tested.[404] In the interim, it is offered to a broader audience of more international scholars in the hope that it might resonate somewhere. The outbreak of xenophobic violence in 2008 in South Africa hangs like a ghost over the conscience of its citizens, who constantly ask whether the critical need for resource allocation (specifically, reallocation) is being dealt with appropriately by a government whose stock answer when faced with any criticism about failed service delivery is lack of capacity. The human health issue associated with mine water decant,[405] the endocrine-disrupting chemical problem that is manifesting as babies being born with both male and female genitalia[406] and men manifesting with lower semen viability,[407] and the whole issue of people with compromised immune systems being exposed to chronic doses of microcystin toxins found in highly eutrophic water in South Africa[408] is thus worthy of being monitored by the international scientific community, because it is likely to yield rich data in the near future.

PART II

EMERGING TRANSCONTINENTAL ISSUES

CONCEPTUALIZATIONS AND CASE STUDIES

CHAPTER 5

ENVIRONMENTAL SECURITY AND NEGLECTED TROPICAL DISEASES IN AFRICA

Chad Briggs and Jennifer Bath

INTRODUCTION

Despite the general agreement of the importance of addressing disease burdens and a general awareness that diseases are related to environmental conditions and to security, there remain significant differences in understanding how disease is an environmental security issue and in responding to such conditions. This chapter lays out a conceptual framework for assessing security risks of tropical disease in Africa, based upon vulnerability models of resilience and fragility used both in ecology and some areas of disaster research. In particular, we focus on the category of "neglected tropical diseases" (NTDs), and in regions of the eastern Sahara and Zimbabwe, illustrate how the relationship between disease and environmental security is complex, systemic, and treatable. By focusing on two diseases, cholera (*Vibrio cholerae*) and Dracunculiasis (*Dracunculus medinensis*), also known as Guinea worm disease, that rely upon availability of clean water, we demonstrate that environmental factors are not the root causes of disease and suffering but dynamic processes where health and security closely interact. Diseases can result from violent conflict and disasters and may act as signals to a larger failure of social, economic, and political systems prior to violence erupting. Diseases also act as an effective point of intervention, where treatment may increase adaptive capacity of the system.

BACKGROUND

The burden of tropical diseases upon populations in Africa is substantial. The category of diseases endemic in warmer climates infects a large proportion of the population worldwide, with upwards of one-third of the world's population infected with one or more diseases. Approximately 350 to 500 million people are reportedly infected with malaria, and up to 2 billion people are infected with tuberculosis (TB), while approximately 2 billion people suffer from helminthiasis (intestinal parasitic worms).[409] Many of the infected individuals harbor more than one disease simultaneously, and such diseases can substantially impact individual health, economic productivity, education levels, and demographic distribution. By comparison, the worldwide infection rate for HIV/AIDS is 33 million, with the majority of such cases existing in sub-Saharan Africa.[410] The reemergence of certain tropical diseases in places not considered endemic has raised questions about the possibility of climate-related spread of disease (e.g., malaria), of globalization spreading more virulent zoonotic diseases (H5N1 influenza or SARS), or of increasing bacteriological resistance to treatment (TB). The primary question is what role the environment plays in the spread and control of disease and what consequences this has for understanding security in Africa.

The historical connection between disease and security was a fairly straightforward concern with disease impacts on troops operating in combat conditions. Prior to the First World War, more soldiers typically died of disease than of direct combat wounds, and Crosby argues that earlier European crusades failed at least in part due to unfamiliarity with and vulnerability to tropical diseases such as malaria.[411] The stability of European colonial rule in tropical regions was potentially hampered by susceptibility to new or imported diseases, and a certain amount of earlier geographical research was devoted to questions of adapting colonial settlers and officers to new climatic conditions.[412] The advent of modern medicines and changes in military operations lessened such concerns considerably in the later twentieth century, and disease did not resurface as a security concern again until the end of the Cold War. At that time, the concept of "emerging" or "reemerging" diseases was developed, with growing concern that changes in environmental conditions in Africa would precipitate the development and spread of new, lethal diseases.[413] The 1990s public concern in the United States over the Ebola virus was based in large part upon historic concerns over dangerous, exotic, tropical diseases.

Since that time, there have been two primary ways to understand the connections between disease and security. The first approach is largely institutional and relies upon Cold War models of interstate violence to understand causality and measure potential impacts. The institutional approach defines security as a function of either interstate violence or the stability of state institutions and holds much in common with the so-called Toronto school approach to environmental security.[414] According to this model, disease affects societies through decreased economic productivity, and this economic decline can trigger or contribute to instability at the state level. Certain diseases such as HIV/AIDS can also hamper military readiness, creating regional imbalances in state power and potential opportunities for adversaries. Environmental conditions in such models are viewed as external variables, where factors such as exposure to zoonotic diseases or the existence of population pressures would act as outside influences.[415] The institutional approach contains potential problems with lines of causality, as violent conflicts more often contribute to disease than vice versa,[416] a condition examined later in reference to Sudan and the Darfur conflict. Regional power imbalances in Africa often occur more at the substate level than internationally, and little evidence exists that diseases such as HIV/AIDS affect militaries in Africa to a significant degree.[417] The concept of stability of state institutions is highly problematic in Africa, particularly when disease outbreaks occur precisely as a result of actions taken by a state to bolster its stability. The cholera epidemic in Zimbabwe is just such an example, where disease and poor public health impact a country while the state itself remains relatively stable. When instability at varying levels does exist, it is rarely caused by environmental conditions or disease, but rather, we should conceive of disease as a marker or medium through which other processes are manifested.

The alternative approach is generally known as "human security" and often refers to protection of individuals from harm or the threat of harm. Developed both as an analytical tool, to distinguish it from state-based theories of security, and as a normative concept for protection of vulnerable populations, human security is often referred to in disease and security analyses.[418] From this perspective, the security of individuals can be threatened by the existence and spread of diseases, measured in terms of excess morbidity and mortality and related impacts. In 2003 the RAND Corporation published a report on disease and security,[419] but despite claiming to adopt a human security approach, it was still concerned in large part with economic impacts to states. A primary focus on the individual may reinforce the microeconomic approach prevalent in much of social science, where environmental concerns (including disease) are external to the person and each is affected autonomously. Application of a human security framework to complex systems runs the methodological risk of considering society as a collective aggregation of individuals, positing that if enough individuals are infected with a major disease, then security might suffer from lack of military readiness or substantial economic harm. The definition then loops back into a traditional version of national security and may miss crucial dynamics contributing to both the problems and potential solutions.

Understanding the connections between these disease burdens and environmental security requires security concepts that move in an intermediary sphere between national and human security and use more systemic assessments of interconnections.[420] An improved approach to security and risk consequences of disease must look at the group or community level, a basic tenet of epidemiology since the 1800s, for the nature of infectious disease is highly social as well as environmental. The nature of communicable diseases depends in part on the natural environment but often can only be explained in terms of human relations and unintended consequences of collective actions. Even when analyses employ the language of systems theories and nonlinear stability shifts, a continued focus on the state and inability to identify mechanisms hampers their applicability. State institutions in Africa may be or become unstable for any number of reasons, while communities have often developed practices to remain resilient against particular risks. State actions, including in the form of violent conflicts, can disrupt the networks and practices at varying levels, thus increasing vulnerability to disease and environmental change. The following section details the concept of vulnerability as it applies to environmental security issues, where the environment is understood as a process for exposing underlying vulnerabilities, not existing as an external, root cause.

VULNERABILITY

Vulnerability does not stand alone as a concept but is rather an aggregate measure of factors influencing risk and the ability to maintain a measure of stability. In complex systems, some of these factors are emergent, meaning that their nature relies upon networks of relationships among components rather than the sum total of aggregate individuals. The concept of vulnerability is drawn both from disaster studies[421] and political economy studies of livelihoods,[422] combined with models of complex systems taken from ecological sciences.[423] The first component in assessing vulnerability is risk, measured as a function of a probabilistic hazard (H) and exposure to the hazard (E), as $R = f(H, E)$. The adverse effects of diseases and people's exposure to such pathogens is widely accepted as a measure of risk. Disaster studies began including "vulnerability," known as the basic hazard function of risk, as a mitigative or multiplier factor in determining specific risks. There are several such factors, and they can be applied in a variety of contexts.[424]

The first such additional factor is sensitivity, or the extent to which an adverse outcome will "push" one away from a baseline measure. Populations that are highly sensitive to disease are those most likely to suffer relatively worse symptoms from the infection. With cholera, for example, children and the elderly are most sensitive and most likely to suffer the worst consequences from the disease. Other diseases, for reasons of age, gender, genetic background, or past health issues, may be more or less sensitive to certain risks than others. Sensitivity can also be used to describe sensitivity to natural disasters (e.g., landslides due to geographic location), economic conditions (e.g., occupations sensitive to changing economic output), or environmental change (e.g., reliance on agriculture sensitive to drought). The second mitigative factor is resilience, or the measure of how effectively an individual or a system can return to its baseline after disruption. The resilience of a system depends upon support networks, whether this is measured in terms of access to vital resources, the ability to depend upon charity and assistance from neighbors, intervention from outside actors, or other characteristics that can contribute to a robust network of interlinking components. Resilience may not translate to a return to the same exact condition, as complex systems may find multiple stability points, a characteristic of observed ecological systems. The last component is fragility, a concept taken from engineering studies but which can be applied to social and ecological systems when not only is a nonlinear shift in stability observed, but the nature of the system itself is significantly compromised and no longer exhibits the same relationships.

Taking these factors together, vulnerability provides an alternative definition for security, when violence may not be present but vital systems may be at significant risk. One aspect of such a perspective is that vulnerability assessments are meant to be scalable, meaning that systems can and should be examined from multiple levels, with measurable components shifting focus at each level as well. Such approaches are already common in epidemiology, where country-level studies would produce too many ecological fallacies and miss crucial local effects and causes. The other importance of vulnerability assessments is the forward nature that such studies can take. Although the examples in this chapter are retrospective, it is possible to use the same tools of analysis as risk assessments, or assessments of those areas most at risk of hazards, as measured by the above components. Such foresight is crucial for security, as those measurements of violence or state stability can be observed only when the situation is often too late for effective action. If traditional models of environment/disease and conflict do not have predictive capacity, risk assessment scenarios and vulnerability assessments are perhaps more suitable for analysis of such dynamic systems. Finally, vulnerability assessments require interdisciplinary analysis, as the various systems at risk (social, political, economic, ecological, epidemiological) are overlapping and closely interlinked. It is not possible to understand fully the operation of one system without the others.

Sudan

Guinea worm disease is the most thoroughly documented parasitic disease in history. Early documentation of the disease appears in an Egyptian papyrus from 1500 BC or earlier[425] that describes the infection in detail. The presence of this worm in ancient Egyptian times is also supported in the findings of a well-preserved male *Dracunculus* worm in the mummy of a young Egyptian woman.[426]

The exact historical timeline of dracunculiasis in Sudan is not known; however, there is significant evidence to suggest that *Dracunculus medinensis* has been endemic in this region for thousands of years. Evidence originates from the parasite being carried by individuals traveling north along trade and pilgrimage routes from Sudan into Egypt prior to 3100 BC,[427] to documentation demonstrating infected slaves from Sudan falling ill at a later date after arriving in Egypt.[428]

The adult *Dracunculus* parasites live in the subcutaneous connective tissues of their host, and after reaching adulthood, the females emerge from the infected individual's skin to release larvae into water. The larvae are ingested by a microscopic freshwater copepod, better known as a water flea. Humans ingest the infected water fleas when they drink from contaminated water sources. The larvae grow up to a meter long as they develop within the human host. Upon maturity of the female worm, the formation of a large ulcer is induced on a lower extremity of the individual, through which the female worm emerges. Emergence of the worm is extremely painful, and relief of pain is often sought by the immersion of the wound into water, usually the community's only source of drinking water. The parasite protrudes from the skin and the *Dracunculus* worm then releases thousands of larvae into the water supply. The larvae are ingested by *water fleas, and the* life cycle continues when individuals drink the larvae-infested water.

Because dracunculiasis is contracted through the drinking of contaminated water, it is entirely preventable. Control of the disease requires access to an uncontaminated water source, prevention of individuals with *Dracunculus*-induced ulcers from entering water sources used for drinking, use of a larvicide to kill water fleas, or filtration of the drinking water through a finely woven cloth. In addition, several features of dracunculiasis make it a good target for eradication. Diagnosis of the disease in unambiguous; the intermediate vector is not mobile; the incubation period within the host and the intermediate vector is of a limited duration; interventions are effective, low-cost, and relatively simple to implement; the disease is limited geographically and seasonally; and there is no known animal reservoir.[429]

The initiative to eradicate dracunculiasis began at the U.S. Centers for Disease Control and Prevention in 1980. In 1991 the disease was still endemic in twenty countries. Over the next several years, the number of cases reported to WHO decreased by 75 percent, from approximately 547,575 cases in 1991 to 130,000 cases in 1995.[430] At this time, however, more than 50 percent of the reported cases every year were coming out of Sudan alone. In 1995 former president Jimmy Carter negotiated a four-month cease-fire agreement in Sudan to allow health-care workers to implement strategies for eradication, including containment of active cases and education about methods to avoid ingestion of contaminated drinking water. These programs relied on the successful training of village-based health workers. The cease-fire allowed relief workers access to almost two thousand endemic villages, as well as the distribution of cloth filters for water decontamination. Unfortunately, a surge of violence related to the civil war halted access to the dracunculiasis-affected villages, particularly in the southern part of the country.

The war also played a critical role in disrupting eradication efforts in Sudan in 2001. At this time, eradication efforts were under way throughout Sudan, with the distribution of 850,000 filters for household use and 7.8 million portable pipe filters to southern Sudan by the Carter center to filter potentially contaminated drinking water.[431] Toward the end of this same year, violence in Sudan resulted in forced evacuations, flight bans, bombings, and the withdrawal of some NGOs due to a dispute with rebel forces over the signing of Memorandum of Understanding. As a result, Sudan reported 78 percent of all cases in 2001, and virtually all of these cases were in the southern states, where political unrest limited the access to endemic regions.[432]

As a direct result of the war in Sudan, access to the regions where dracunculiasis is endemic has been limited, making it impossible to implement and maintain an effective eradication program. As Sudan has struggled with conflicts in the western Darfur region and, previously, in its oil-rich south, the breakdown in the economy and the health of the Sudanese is directly reflected in the number of dracunculiasis cases still reported out of Sudan. The strategic efforts of the Carter Center, the Centers for Disease Control and Prevention (CDC), the United Nations Children's Fund (UNICEF), and the World Health Organization (WHO) have reduced *Dracunculus* infections to record lows in many countries, and cases are now reported in only six countries in sub-Saharan Africa.[433] In 2007, 9,585 cases were reported, but a staggering 61 percent of the cases were reported from Sudan alone.[434] In January through June of 2008, 98 percent of the cases were reported from Sudan, Ghana, and Mali. Sporadic violence, in particular in Sudan, will continue to thwart health relief efforts, making eradication impossible until such a time that the violence is brought to a halt. The continued levels of dracunculia-sis in Sudan pose not only a major health concern to the people of this region but a threat to the success of the global dracunculiasis eradication program as well.

ZIMBABWE

Cholera was originally associated with waters of the Ganges River delta in India, where the *Vibrio cholerae* bacterium was endemic. It did not spread globally until the early 1800s, when trade routes and increased urbanization increased both range and transmissibility, and the first of several pandemics hit between 1816 and 1826.[435] A diarrheal disease associated with drinking water contaminated with human waste, those infected with cholera suffer rapid dehydration caused by bacterial secre-tion of an enterotoxin in the small intestine. Cholera can exhibit rapid onset of symptoms, as those infected can become fatally overcome in a matter of two to three hours without treatment. Infections are more typi-cally fatal after eighteen hours to several days, and untreated cholera can be fatal in over 10 percent of cases, although simple treatments can keep this rate down to 1 percent.[436] Effective treatment requires rehydration, including sugars and salts, but, of course, must ensure that the rehydrat-ing fluids are themselves not infected. Developments in water filtering, chlorination, and urban planning have effectively controlled the disease in many parts of the globe, although it remains endemic in areas without access to clean drinking water.

Cholera has historically been endemic in Zimbabwe, with cases reported annually since 1992. Those cases were often considered isolated and sometimes linked to neighboring countries, while within Zimbabwe, rapid advances in health care, public health policies, and environmental conditions resulted in significant increases in life expectancy through the 1980s. Yet, compared to 1994, the life expectancy of women had by 2008 dropped from fifty-four years to thirty-seven, and infant mortal-ity rates had risen significantly.[437] The 2008 cholera outbreak in Zimba-bwe, which by the end of 2008 had resulted in an estimated 60,000-plus

infections and several thousand reported deaths, was the result of particular policies combining with environmental conditions, made worse
by a general denial of the situation until toward the end of 2008.[438] The
epidemic cannot be explained by reference to any outside events, nor did
the outbreak occur spontaneously, but is rather understood as a symptom
of systemic breakdowns within the country. Cholera is therefore not a
cause of instability but a marker of and a feedback to instability in the
economic, health, and environmental systems of the country. The government's slow response to admitting the existence of a problem and the
subsequent blame placed on outside influences only served to worsen a
rapidly deteriorating situation.

The roots of the cholera epidemic can be traced back to seizure of
farmland from predominantly white farmers in 2000, itself a response
to racial tensions and land ownership patterns that date to the establishment of Rhodesia in 1965. The land seizures, which prompted economic
sanctions from the United Kingdom and heavily disrupted agricultural
production in Zimbabwe, led to the collapse of the agricultural industry
in the country, a primary source of export income. Annual wheat production dropped from 300,000 tons in 1990 to less than 50,000 tons in
2007. Tobacco production, which had accounted for more than a third of
Zimbabwe's foreign exchange, by 2007 had dropped to only one-quarter
of its 2000 production levels.[439] The economic system, tied closely to
cash crop exports, was highly sensitive to any disruption and, in retro-
spect, was highly fragile when the dislocations became large enough.
The loss of agricultural production, combined with corrupt government
practices, led to massive infusions of cash into the economy that touched
off the current hyperinflation of the Zimbabwean dollar. By early 2009,
the Zimbabwean dollar was effectively worthless and had been replaced
in practice by the U.S. dollar and South African rand.

The economic crisis in Zimbabwe had cascading effects elsewhere
and was made worse by government policies to destroy the "informal
economy" in the country, meaning those sectors not dependent on currency payments. The 2005 "Operation Murambatsvina" was aimed at
driving out more than 700,000 residents of impoverished urban areas, a

potential source of unrest that the rivals to Mugabe's government might have relied upon.[440] As a result, livelihood networks were heavily disrupted at the same time that spiraling currency led to underinvestment in and underpayment of water treatment facilities and medical resources. General vulnerability was increased as access to social and financial resources was deliberately or inadvertently destroyed. By early 2008, many areas had lost access to clean drinking water, prompting residents to dig shallow wells for water. But, absent working plumbing and proper planning for such wells, they were easily infected with spillage of human waste, providing a transmission mechanism for the cholera bacteria.

Transmission of cholera does not, in itself, lead to a disease outbreak of epidemic proportions. But the environmental and economic conditions combined to form positive feedback loops for ever-widening outbreaks of the disease and the likelihood that the disease would go untreated. Because water from municipal sources began disappearing at roughly the same time around the country, and as Zimbabwe's geography tends toward high water tables, easily contaminated shallow wells created broad conditions for outbreaks, not merely point-source contaminations. The state-run health-care service also ceased to function by late 2008, the result of inflationary currency rendering paychecks meaningless and basic upkeep for hospitals impossible, despite the best efforts of physicians and nurses trying to care for cholera patients. Cholera had the added effect of draining resources from an already overtaxed health system, forcing health centers to refuse treatment not only to cholera patients but also to those being treated for other diseases, such as HIV/AIDS and malaria.[441] Zimbabwe's relatively high HIV/AIDS infection rates created greater sensitivity among those afflicted with cholera and, therefore, higher vulnerability in the system as a whole. The cholera epidemic worsened after December 2008, as the spread of the disease coincided with onset of the rainy season and greater movement of peoples during the Christmas holidays.[442] The loss of agricultural production and hyperinflation has also caused a large-scale food shortage in the country, with over half of Zimbabwe's population dependent upon international aid shipments of food. Such food insecurity increasing sensitivity to disease prompts increased migration among certain segments of the population and, there-fore, ensures greater transmission of cholera from one area to another.[443]

In a vulnerability framework, consideration of cholera's impacts is only one aspect of the overall assessment. Economic conditions in the coun-try touched off a series of cascading events, and the political responses ensured a positive feedback loop that allowed the greater spread of chol-era. In a robust system, responses to an outbreak of cholera would be twofold: an immediate response to isolate the disease and remove the potential risks of the disease spreading, and a retrospective analysis to identify what failure allowed cholera to be introduced to the environ-ment/health system. The existence of a disease does not mean the intro-duction of instability pressure per se; that depends upon the nature of the system and its responses.[444] Do public health authorities recognize the cause and work to correct the situation? Is maintenance of water treat-ment systems supported? Do people have access to adequate resources (medical, food)?

CONCLUSION

Examining events such as the cholera outbreak in Zimbabwe or the near-eradication of guinea worm in East Africa, one can trace patterns of vul-nerability and response and perhaps use such approaches to understand where African societies may be vulnerable in the future. The concept of stability as a function of the state alone is an unwieldy unit of analysis, forcing aggregation of complex data and interrelationships. The useful-ness of analyzing security from the perspective of infectious disease lies not in any expectation that diseases will cause the breakup of states. Even under the worst pandemics of influenza in 1919, little evidence exists that this was ever a danger. Rather, diseases can act both as mark-ers for wider instability and as potential ways to increase adaptive capac-ity from the ground up.

In the case of Zimbabwe, tracing events back from the cholera out-break reveals a systematic breakdown of infrastructure and support in the country, caused, in large part, by deliberate attacks on rivals and their means of support. This form of resilience targeting is common in civil conflicts, where natural resources become the means for damaging one's opponents.[445] Cholera serves not only as an analytical marker for state failure and violence but as an important psychological component of trust in the state. A major challenge of Prime Minister Morgan Tsvangirai will be control of the disease, as evidence that overall conditions will improve. Such improvements in conditions were demonstrated with dracunculiasis, where a major effort started by the U.S. government has managed to control a formerly debilitating disease in East Africa. Despite the ongoing conflict in Sudan hampering efforts to eradicate the disease, the overall benefits of dracunculiasis-free communities are significant in relation to the costs of control and increase the ability of groups to adapt to other adverse conditions and challenges. The requisite focus on clean water as a function of control policies also has secondary benefits to health and well-being, which themselves tie into the United Nations' Millennium Development Goals for less-developed regions.

The complex systems that underlie environmental security require greater understanding of how environment, society, economies, political structures, and military operations interact. The feedback loops and cascading effects (both positive and negative) of actions are perhaps even more crucial when such systems are fragile, highly sensitive to change, or not resilient enough to stabilize after disruptive events. We cannot rely solely on security concepts that must wait until the outbreak of violence or collapse of a political system before acting. To do so would prevent effective intervention by the international community. Rather, effective foresight assessments for the environment in Africa may make use of existing methods in risk and public health to understand better the stability of substate systems and where effective intervention can be made to create cascades of positive effects in communities.

CHAPTER 6

THE MOST VULNERABLE CONTINENT

AFRICA AND GLOBAL CLIMATE CHANGE

John T. Ackerman[446]

Global climate change is predicted to have an enormous impact on the continent of Africa. Currently, Africans are pressured by a variety of economic, social, political, technological, and environmental challenges that will only be exacerbated by global warming:

> Deterioration in terms of trade, inappropriate policies, high population growth rates, and lack of significant investment—coupled with a highly variable climate—have made it difficult for several countries to develop patterns of livelihood that would reduce pressure on the natural resource base. Under the assumption that access to adequate financing is not provided, Africa is the continent most

vulnerable to the impacts of projected changes because widespread poverty limits adaptation capabilities.[447]

INTRODUCTION

Climate change projections indicate that by 2020, between 75 and 250 million Africans will be exposed to increased water stress due to climate change.[448] Also by 2020, yields in African states that are dependent on rain-fed agriculture could be reduced by up to 50 percent. Most importantly, agricultural production, including access to food, in many African countries is projected to be severely compromised. This would further adversely affect food security and exacerbate malnutrition.[449] In addition, by 2080, the amount of arid and semiarid land is projected to increase 5 to 8 percent under a range of climate scenarios.[450] Finally, low-lying African coastal communities will be severely stressed by projected sea-level rise, and the cost of adaptation could amount to at least 5 to 10 percent of overall GDP.[451] Clearly, Africa is the continent most at risk from global climate change; however, what, specifically, does Africa's vulnerability entail?

Analyzing the risks associated with global climate change requires data and fact-based empirical evidence if the specific vulnerabilities are to be exposed and effectively addressed. In addition, the vulnerabilities are influenced by certain factors that can be examined rigorously, and these factors include a state's exposure to climate stimuli, the state's sensitivity to climate change, and the mitigation/adaptation capacity of the state. Together, these factors will determine the social, political, economic, and environmental systems' vulnerability to climate change. Distinctively, this research examines the environmental vulnerabilities of select African states to "get inside the numbers" and reveal particular aspects and facets of this very consequential state characteristic.

Environmental vulnerability is ultimately impacted by the environmental health and ecosystem vitality policies a state legislates. For example, a state's agricultural policies will influence carbon-emission mitigation in this sector. Also, the water-management policies a state crafts will determine the water system's sensitivity to drought, and state policies on climate-influenced disease vectors directly affect exposure levels. The process of getting inside the numbers and analyzing the often-obscured endogenous data and performance measures enables "identification of leaders and laggards, highlights best policy practices for each issue, and identifies priorities for action for each country."[452] A goal of the research is to provide information for the laggards to look to the leaders for help on how to reduce their environmental vulnerability to climate change.[453]

Evaluation of the environmental vulnerabilities of African states to climate change necessitates an examination of key background material. First, a brief overview of some of the general impacts climate change scientists are predicting on the African continent is needed to establish context. Second, the criteria and indicator categories for evaluating environmental vulnerability to climate change must be defined and explained. Once the background material is provided, then the evaluation process can be applied to four cases. In this instance, a single state from North Africa, south Africa, the Great Lakes Region (east Africa), and west Africa, respectively, was chosen for deeper analysis. Each case will be evaluated for exposure, sensitivity, and adaptation/mitigation capacity as related to environmental vulnerabilities to climate change. After analysis of each state, case-specific conclusions will be drawn and recommendations offered.

BACKGROUND: EXPECTED CLIMATE CHANGE IMPACTS ON AFRICA

The continent of Africa is extremely vulnerable to changes in climate. Interaction of a variety of factors at the continental, regional, and local levels places enormous stress on the environmental, governance, and institutional components of all African states. Continent-wide endemic poverty, corruption, limited capital, fragile markets, technological retardation, and ecosystem degradation pressure concurrent weak adaptation and mitigation capacities and increase exposure and sensitivity levels.[454] Regionally, the six distinct African climatic zones—tropical wet, tropical summer rainfall, semiarid, arid, highland, and Mediterranean—are all experiencing variability that is driving detectable changes in ecosystems. For example, grasslands in semiarid regions have been subjected to measurable deceases in rainfall, and marine ecosystems in the Mediterranean zone have experienced increases in average annual sea-surface temperatures. In addition, regional changes in water availability, accessibility, and demand will aggravate water stress challenges in several African states. Coastal regions will also face challenges created by climate change. Already-threatened mangroves could be adversely affected by salt water intrusion, and coral ecosystems could decline as a result of increasing water temperatures. Sea-level rise alone could cost coastal Africans at least 5 to 10 percent of their gross domestic product in adaptation responses. Locally, most African farmers are dependent on seasonal rains for crop production, and some African states are predicted to suffer from reductions in crop yields as large as 50 percent by 2020 as a result of climate changes. Finally, climate change may change the spatial and temporal relationships between infectious disease vectors and victims and enhance the spread of diseases such as dengue fever, malaria, and cholera.[455]

Environmental Vulnerability Model and Criteria

The conceptual basis for vulnerability is grounded on a framework similar to one developed by Hans-Martin Fussel and Richard J. T. Klein.[456] This framework was integrated with another framework developed by researchers at Yale and Columbia universities.[457] The components of the integrated model are defined below, and a graphical depiction of the model is also provided for further clarification. The model has been modified to evaluate certain environmental vulnerability criteria.

Model Components
Adaptation
Changes in policies and practices designed to deal with climate threats and risks. Adaptation can refer to changes that protect livelihoods, prevent loss of lives, or protect economic assets and the

environment. Examples include changing agricultural crops to deal
with changing seasons and weather patterns, increasing water conservation to deal with changing rainfall levels, and developing medicines and preventive behaviors to deal with spreading diseases.[458]

Climate Change

"Climate change refers to any change in climate over time, whether due
to natural variability or as a result of human activity."[459]

Exposure

"The nature of degree to which a system is exposed to significant climatic variations."[460]

Greenhouse Gas

Greenhouse gases are those gaseous constituents of the atmosphere, both natural and anthropogenic, that absorb and emit
radiation at specific wavelengths within the spectrum of infra-
red radiation emitted by the Earth's surface, the atmosphere, and
clouds. This property causes the greenhouse effect. Water vapor
(H_2O), carbon dioxide (CO_2), nitrous oxide (N_2O), methane (CH_4)
and ozone (O_3) are the primary greenhouse gases in the Earth's
atmosphere. As well as CO_2, N_2O, and CH_4, the Kyoto Protocol deals with the greenhouse gases sulphur hexafluoride (SF_6),
hydrofluorocarbons (HFCs), and perfluorocarbons (PFCs).[461]

Greenhouse Gas Emissions

Direct emissions or "point of emission" are defined at the point
in the energy chain where they are released and are attributed
to that point in the energy chain, whether a sector, a technology
or an activity. E.g. emissions from coal-fired power plants are
considered direct emissions from the energy supply sector. Indirect emissions or emissions "allocated to the end-use sector" refer
to the energy use in end-use sectors and account for the emissions
associated with the upstream production of the end-use energy.
E.g. some emissions associated with electricity generation can be
attributed to the buildings sector corresponding to the building
sector's use of electricity.[462]

Impacts

"The consequences of climate change on natural and human systems."[463]
The consequences can be distinguished by biophysical (natural) impacts
and social (human) impacts as well. This chapter concentrates on the
biophysical, or environmental, impacts.

Mitigation

"An anthropogenic intervention to reduce the anthropogenic forcing
of the climate system; it includes strategies to reduce greenhouse gas
sources and emissions and enhancing greenhouse gas sinks."[464]

Sensitivity

> Sensitivity is the degree to which a system is affected, either
> adversely or beneficially, by climate variability or change. The
> effect may be direct (e.g., a change in crop yield in response to a
> change in the mean, range or variability of temperature) or indirect
> (e.g., damages caused by an increase in the frequency of coastal
> flooding due to sea-level rise).[465]

Vulnerability

> Vulnerability is the degree to which a system is susceptible to, and
> unable to cope with, adverse effects of climate change, including
> climate variability and extremes. Vulnerability is a function of
> the character, magnitude, and rate of climate change and varia-
> tion to which a system is exposed, its sensitivity, and its adaptive
> capacity.[466]

This chapter will concentrate on the environmental vulnerability to climate change.

DATA AND EVALUATION CRITERIA

Quantitative and qualitative environmental data were used to evaluate
the differential vulnerabilities to climate change of selected African
states. As noted before, the differential vulnerabilities are "a function
of the character, magnitude, and rate of climate change and variation to

which a system is exposed, its sensitivity, and its adaptive capacity."[467] The criteria for evaluation are the respective exposure and sensitivity levels of each state and its subsequent capacities to mitigate and adapt to climate change. The evaluation criteria will delineate each state's overall environmental vulnerability to global warming.

The quantitative indicators used for the case evaluations come from Yale and Columbia universities' Environmental Performance Index (EPI).[468] Specifically,

> the 2008 EPI deploys a proximity-to-target methodology, which quantitatively tracks national performance on a core set of environmental policy goals for which every government can be—and should be—held accountable. By identifying specific targets and measuring the distance between the target and current national achievement, the EPI provides both an empirical foundation for policy analysis and a context for evaluating performance.[469]

The EPI performance indicators are based on two broad environmental objectives that reside in and overlap every state's biophysical domain: environmental health and ecosystem vitality. The environmental health objective is to reduce the environmental stresses on human health.[470] The ecosystem vitality objective is to diminish loss and degradation of ecosystems and natural resources.[471] The two overarching objectives are further broken down into six policy categories and ten more issue-focused policy performance indicators. The ten policy performance measures that will be discussed are environmental burden of disease, water (effects on humans), air pollution (effects on humans), air pollution (effects on ecosystems), water (effects on ecosystems), biodiversity and habitat, forestry, fisheries, agriculture, and climate change.[472] The EPI is the best environmental data available, yet it does contain some gaps and weaknesses. In addition, the current ranking system only covers 149 out of 238 possible states, and data are lacking in climate-related areas such as wetlands loss, freshwater ecosystem health, and agricultural soil quality.[473] The qualitative data are derived primarily from the *Climate Change 2007: Impacts, Adaptation, and Vulnerability* report by

Working Group II of the Intergovernmental Panel on Climate Change (IPCC).[474] Also, the United Nations Environment Programme's *Global Environment Outlook: GEO 4* report provided substantial background information on Africa's biophysical characteristics.[475]

As stated before, the initial objective of the evaluation is to reveal how the ten indicators impact the exposure/sensitivity levels and mitigation/adaptation capacities of the selected states to global climate change. Ultimately, quantitative and qualitative analysis of the measures provides refined insight into the specific vulnerabilities of these countries and may expose "an entry point to more detailed explorations"[476] of the global climate change threat.

POLICY INDICATORS

A brief description of the ten policy indicators, their linkages to African states, and their relationship to climate change is provided in the following sections. The correlation between the policy indicator and the vulnerability to climate change criteria is not perfect in each category, and there is overlap between criteria and indicators. Nevertheless, the indicators do provide insight into global exposure/sensitivity levels and mitigation/adaptation activities.

Each policy indicator is supported by widely used and respected data sources. The environmental burden of disease indicator comes from the World Health Organization (WHO) and is broadly regarded to be the most comprehensive and carefully defined measure of environmental health burdens. Additional data for the environmental health indicators come from the United Nations Environment Programme (UNEP), World Health Organization (WHO), and United Nations Children's Fund (UNICEF) and are widely accepted as important indicators of the burden of water and air pollution on human health. The ecosystem vitality data come from various sources, including the Nature Conservancy, the UN's Food and Agriculture Organization (FAO), the University of British Columbia's Fisheries Centre, the Center for Earth Science Information Network (CIESIN), UNEP, and the International Energy Agency (IEA).[477]

FIGURE 6.1. Environmental Vulnerability Model.

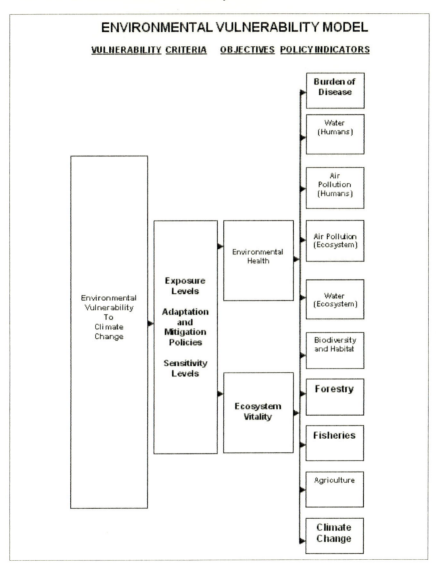

Source. Modified from Daniel C. Esty, M. A. Levy, C. H. Kim, A. de Sherbinin, T. Srebotn-jak, and V. Mara, *2008 Environmental Performance Index* (New Haven, CT: Yale Center for Environmental Law and Policy, 2008), 18.

Environmental Burden of Disease

This indicator helps measure the impact of environmental conditions on human health. Specifically, the data indicate the "number of life years lost due to premature mortality caused by environmentally influenced disease and the years of healthy life lost due to disability caused by such disease."[478] African states are challenged by a range of environmentally influenced pathogens. For example, certain disease vectors in Africa, such as malaria, are predicted to increase their negative impact on human health as a result of climate change.[479] Importantly, the EPI environmental burden of disease indicator provides information on policies that affect exposure and sensitivity to the impacts of global climate change.

Water (Humans)

Clean water is essential to human health, and without adequate water management and sanitation, the potential for contamination by fecal-oral pathogens and waterborne diseases is dramatically increased. For example, waterborne diarrheal diseases are significant causes of death and disability in Africa.[480] As climate changes, the potential for droughts and floods increases, as does the potential for water contamination. Millions of Africans are particularly vulnerable to droughts and floods, exacerbating problems of access to clean water and exposing millions to diarrhea, cholera, and malaria.[481] Overall, the water indicators measure exposure and sensitivity levels to the impact of global warming.

Air Pollution (Humans)

The air pollution indicator measures both indoor and outdoor air pollution. Developing countries rely heavily on biomass (wood, charcoal, dung, and crop residue) for cooking fuel, and these fuels create severe indoor health risks by releasing toxins and greenhouse gases into the air.[482] Africans burn biomass for more than 80 percent of their energy needs, and in some African states (Kenya, Tanzania, Mozambique, and Zambia), wood or charcoal is used by more than 90 percent of households for cooking food.[483] Outdoors, as populations in many African states procure biomass for energy, large areas are cleared of vegetation that could act as carbon sinks. In addition, the outdoor measures are indicators of industrialization and fossil fuel use, both of which release great amounts of greenhouse gases. The measures of air pollution that affect humans can also specify exposure and sensitivity policies towards climate change impacts; however, they may indicate mitigation and adaptation efforts as well.

Air Pollution (Ecosystems)

Air pollution not only affects human health but also has adverse affects on ecosystems. Small reactive compounds such as ozone (O_3), benzene (C_6H_6), sulfur dioxide (SO_2), nitrogen oxides (NOx), and volatile organic compounds (VOCs) have a range of negative environmental impacts. For example, ozone degrades plant cuticles through oxidation, inhibiting plant development and growth. Sulfur dioxide and nitrogen oxides both react with other atmospheric compounds, resulting in acid rain. Prolonged ecosystem exposure to acid rain can diminish fish stocks, decrease biological diversity in acid-sensitive lakes, degrade forests and soils, and diminish agricultural productivity.[484]

Rural African ecosystems are not significantly threatened by air pollution, but urban ecosystems are coming under increasing pressure from industrial sites and fossil fuel energy plants. In addition, increased desertification of some African states in the Sahel region, augmented by climate change, has increased airborne contaminants and is negatively impacting the environmental health and ecosystem vitality of the states in this region.[485] Generally, indicators of air pollution that impinge on ecosystems can identify exposure and sensitivity policies, but there is some overlap with policies that mitigate and enable adaptation to climate change.

Water (Ecosystems)

Clean water is also essential for the growth and development of every ecosystem. Ecosystems themselves contribute enormously to the quality and quantity of clean water available to all living creatures.[486] Climate change is increasing water stress throughout Africa by negatively impacting water usage levels, ease of use, and ease of access. Water stress within African populations is predicted to triple from approximately 200 million people to more than 600 million people by 2050.[487] Policies that ensure adequate quantities of clean water for ecosystems can, in addition, denote policies that reduce exposure and sensitivity to climate change.

Biodiversity and Habitat

Biodiversity, the natural variety and processes of living species, reflects the health of biological systems. Healthy biodiversity and habitats ensure ecosystem services such as soil regeneration, flood control, food production, water/air purification, and climate regulation are balanced and maintained.[488] Ongoing human activities such as deforestation and burning of cropland, grasslands, and forests destroy critical habitat, reduce biodiversity, and release large quantities of greenhouse gases. African biodiversity is varied and abundant; however, human development and climate change threaten to alter and degrade habitats and biodiversity throughout the continent.[489] Efforts to protect and conserve biodiversity and habitats reflect both exposure and sensitivity policies as well as mitigation and adaptation programs.

Forestry

Forests are an example of an essential habitat that is a haven for biodiversity, an engine for ecosystem services, a renewable source of natural goods, and a major global carbon sink. Deforestation releases approximately 20 percent of global greenhouse gas emissions, while reforestation can largely offset the emissions, restore critical habitat, and provide sustainable resources and jobs.[490] African forests are under severe threat from deforestation, burning/clearing for farmland, fuel wood extraction, desertification, and climate change. Of particular note is that 70 percent of all detected forest fires are in the tropics, 50 percent of these fires were observed in Africa, and more than half of all forested areas in Africa were burned in 2000.[491] The EPI measures of effective forestry policies essentially reflect mitigation and adaptation policies but do display elements of policies that decease exposure and sensitivity to the impacts of climate change.

Fisheries

The state of fisheries around the world is dire: "Over 70% of all fisheries are over-exploited or fished to capacity. At the current rate of exploitation, most are predicted to collapse by midcentury."[492] Global fisheries provide a significant percentage of the affordable protein in human diets and are a major employer worldwide. One pathway by which global warming is predicted to affect African fisheries is the negative impact rising sea levels are having on mangrove ecosystems and the bleaching of coral reefs as a result of increased sea temperatures.[493] Another pathway is salt water intrusion into critical lagoons and estuaries that are nurseries for many different valuable species of fish and crustaceans.[494] Primarily, fisheries policies evaluated can supplement reduction of exposure and sensitivity levels to global warming impacts.

Agriculture

The expansion of modern agricultural techniques is having global impacts. As mentioned earlier, deforestation and burning of forests, grasslands, and agricultural fields not only reduce ecosystem services, biodiversity, and habitats, they also release vast quantities of greenhouse gases into the atmosphere. A second feature of modern agriculture is the rapid increase in livestock production, which also releases significant emissions of greenhouse gases. A third agricultural issue is the global demand for freshwater for irrigation that is consuming enormous quantities of water from lakes, rivers, streams, and underground aquifers.[495] As a base for food security, African agriculture is threatened by climate change in various ways. Overall, climate change is expected to reduce growing seasons and eliminate farming on marginal lands.[496] Most African agriculture is by subsistence farmers that rely on rainfall for irrigation and do not have access to technological advances in high-yielding seeds, fertilizers, and pesticides. These disadvantages make African farmers particularly susceptible to climate variability and change.[497] In sum, the agricultural policy measures can be viewed as a mixed bag of exposure and sensitivity efforts and mitigation and adaptation programs.

Climate Change
Measures of climate change reflect the disparate levels of greenhouse gas emissions, energy sources, and overall economic development.

The forecasted impacts of global climate change, from sea-level rise, coastal flooding, and extensive glacial deterioration to droughts, heat waves, and desertification, are already being felt globally and are projected to increase in severity. These events are expected to increasingly affect human health, water resources, agriculture, and ecosystems. While most greenhouse gas (GHG) emissions to date have originated in developed countries, developing countries are already and will continue to be the most significantly impacted by the consequences of climate change.[498]

African states almost uniformly produce low levels of greenhouse gas emissions. Reliance on subsistence farming, biomass for fuels/energy, and human/animal-powered transport systems ensures African per capita emissions are some of the lowest globally. Nevertheless, these activities often "degrade the environment and compound vulnerability" and make most Africans extremely susceptible to climate change.[499] On the whole, the EPI's climate change indicators measure efforts to mitigate or adapt to climate change impacts.

CASE EVALUATIONS

Four African states were selected from four geographic regions of Africa: north, south, east, and west. The states selected are principally the region's most prosperous, populous states and are considered regional leaders. These states were also selected based on the assumption that their environmental vulnerabilities and responses to these vulnerabilities could be representative of the optimum capability in the region. All other states in the region are assumed to have less-than-optimal capacities overall.

North Africa
Egypt was selected for analysis from North Africa. Egypt has one of the highest EPI ratings in all of Africa (76.3); nevertheless, the score is relatively low compared with the highest score, of Switzerland (95.5),

but substantially higher than the lowest overall score, of Niger (39.1). In addition, Egypt has the fifth-highest rating of all African countries (5/43) evaluated.[500] (See table 6.1 for more details). Further examination of the ecosystem vitality and environmental health of Egypt's biophysical domain will reveal more insights into exposure, sensitivity, and mitigation/adaptation capacities to climate change.

Egyptian ecological air pollution policies are slightly less advanced than either states in the same income (95.8) or geographic groups (92.9) but are still, overall, very high (90.1). This may indicate that Egyptian leaders can design and implement more effective air pollution policies by modeling their geographic or income group neighbors. Water policies toward ecosystems were more advanced in Egypt (67.6) than in peer income (63.4) or geographic groups (37.8), indicating that Egyptian leaders are taking this issue seriously, but the low overall score indicates that substantial progress in improving water quality and reducing water stress still needs to be accomplished. In contrast, Egypt's high ranking among desert states (5) indicates some success in dealing with aridity and water management challenges as compared to other peer-group desert states.[501] Egyptian leaders have also made progress protecting biodiversity and habitats compared to income peers (41.5) and neighbors (36.5), and efforts to conserve critical habitat and biodiversity yielded a score of 77.2. Efforts to protect, preserve, and improve natural resources through forestry, fisheries, and agricultural policies appear to be more effective than income peers (78.4) and geographic peers (77.8), and the overall Egyptian score of 82.0 is laudable. On the other hand, a sector that needs improvement is Egyptian policy towards climate change. The score of 68.9 was better than Egypt's neighbors (59.2) but was lower than Egypt's income peers (72.3). Apparently, Egypt is making more progress than its neighbors but is not as accomplished as states with similar incomes.[502]

The environmental health composite score indicated that Egypt (79.6) is lagging behind peer income (83.2) and geographic groups (82.9) but is still performing well. Conceivably, peer states may be able to help reduce Egyptian exposure and sensitivity to climate change impacts even

TABLE **6.1.** EPI scores for African states.[503]

Rank	Country	Score	Rank	Country	Score
1/58	Mauritius	78.1	23/123	Ethiopia	58.8
2/59	Tunisia	78.1	**24/126**	**Nigeria**	**56.2**
3/64	Gabon	77.3	25/127	Benin	56.1
4/66	Algeria	77.0	26/128	Centr. Afr. Rep.	56.0
5/71	**Egypt**	**76.3**	27/129	Sudan	55.5
6/82	Morocco	72.1	28/130	Zambia	55.1
7/86	Ghana	70.8	29/131	Rwanda	54.9
8/88	Namibia	70.6	30/132	Burundi	54.7
9/93	Congo	69.7	31/133	Madagascar	54.6
10/95	Zimbabwe	69.3	32/134	Mozambique	53.9
11/96	**Kenya**	**69.0**	33/138	Guinea	51.3
12/97	**South Africa**	**69.0**	34/139	Djibouti	50.5
13/98	Botswana	68.7	35/140	Guinea-Bissau	49.7
14/103	Côte d'Ivoire	65.2	36/142	D. R. Congo	47.3
15/113	Tanzania	63.9	37/143	Chad	45.9
16/114	Cameroon	63.8	38/144	Burkina Faso	44.3
17/115	Senegal	62.8	39/145	Mali	44.3

(continued on next page)

TABLE 6.1. *(continued)*

Rank	Country	Score	Rank	Country	Score
18/116	Togo	62.3	40/146	Mauritania	44.2
19/117	Uganda	61.6	41/147	Sierra Leone	40.0
20/118	Swaziland	61.3	42/148	Angola	39.5
21/121	Malawi	59.9	43/149	Niger	39.1
22/122	Eritrea	59.4			

Source. Esty et al., *2008 Environmental Performance Index 2008*, 14.

further with more focused and refined environmental health policies. Finally, EPI's cluster-country analysis placed Egypt in a group of states that performed well on environmental health measures but had relatively low biodiversity scores and relied upon carbon-intensive economies that generate high particulate concentrations of gases such as sulfur dioxide and ozone.[504] The cluster states could all benefit from stronger mitigation and adaptation policies centered on energy generation, transport, and industry. The next process, looking inside the policy scores and evaluating the individual indicator data, reveals even more about Egyptian environmental policies and vulnerability to climate change.

Egyptian efforts to reduce the burden of disease, indoor air pollution, and agricultural subsidies are effective and help decrease exposure and sensitivity levels to climate change. Additionally, while efforts to regulate pesticides and reduce burning of cropland, grasses, and forests are strong, the percentage of land devoted to agriculture is extremely high, and much of this land is becoming degraded.[505] A particular concern is the high levels of urban particulates emissions. Clearly, more mitigation policies are needed here. In addition, efforts to reduce water stress on humans and on agricultural activities are not very effective and are detrimental to reducing exposure and sensitivity to climate change.[506] Studies indicate that Egypt is exploiting almost all of the freshwater resources available and will continue to be burdened by high water-stress levels unless more sustainable water practices are implemented.[507]

West Africa

Nigeria was selected for evaluation from west African states. Nigeria has a very low EPI rating overall (126 out of 149 states evaluated), with a score of 56.2. Also, within the African Union, Nigeria is twenty-third of forty-three states, and within sub-Saharan states, twentieth out of thirty-eight states evaluated.[508] The low overall ranking of Nigeria can be better understood by examining the environmental health and ecosystem vitality scores. Examining these scores will also reveal more about the exposure and sensitivity levels and mitigation and adaptation abilities in countering climate change within the Nigerian state.

The ecological air pollution scores of Nigeria (65.1) were very low, both in comparison to peer income groups (89.9) and geographic groups (89.6). It is clear that Nigeria's efforts to reduce the impacts of climate change from air pollution could be bolstered by examining what income and geographic peers have implemented. Nigerian water policies towards ecosystems are also very weak (57.5). They reflect an overall weakness in the efforts of their peers (58.8), and perhaps outside assistance to better manage water stress and quality would help all states in the west African region. The assistance could lessen the exposure and sensitivity of these west African ecosystems to climate change impacts. Another area in need of outside assistance is the protection of biodiversity and habitats. Low rankings here, individually (59.8) and by peer income (57.8) and geographic groups (62.3), indicate a regional conservation enterprise would be beneficial to mitigation and adaptation capacities. One effective area is the efforts Nigerians are taking towards climate change (85.5), both in comparison to income peers (77.3) and geographic peers (77.2). The climate change policies will directly improve mitigation and adaptation capacities, and sharing of information could help both income and geographic peers effectively prepare for global climate change.[509]

Nigeria' composite environmental health score (40.6) was lower than those of its geographic neighbors (43.0) but was not as low as sister income-group states (32.5). This low score highlights the extreme high levels of exposure and sensitivity to climate change in Nigeria and in the region as a whole. The region will need outside assistance to develop comprehensive measures to reduce exposure and sensitivity, and within Nigeria, certain areas can be targeted, which will be discussed later. One further insight: Nigeria was placed in a country cluster with developing states whose economies are in transition. These states had poor environmental health scores, but because of their transition economies, their carbon intensity levels were low.[510] Detailed examination of some of the individual indicators discloses more about the exposure and sensitivity levels as well as mitigation and adaptation competencies.[511]

The extreme environmental burden of disease, lack of sanitation facilities and clean drinking water, and high levels of indoor air pollution all reduce the resiliency of Nigerians to climate change and make them more susceptible to the negative impacts. In addition, the dearth of biodiversity and habitat conservation efforts combined with high levels of forests converted to cropland expansively reduce mitigation and adaptation capacities while increasing exposure and sensitivity to global warming. Nigerians also need stronger mitigation policies in relation to the emissions of greenhouse gases for electricity. As a net oil exporter, Nigeria is clearly dependent on fossil fuels for income as well as for energy. Mitigation and adaptation efforts that focus more on renewable sources of energy would benefit all Nigerians. On the positive side, Nigeria's emissions per capita and overall industrial carbon intensity are very low and could indicate that transitions to carbon-free/neutral-energy and industrial processes are possible.[512]

South Africa

The state of South Africa was chosen for evaluation from southern African states. South Africa has an EPI score of 69.0, which places it ninety-seventh out of 149 states overall and eleventh out of forty-three African states. In addition, South Africa is ranked eighth out of thirty-eight sub-Saharan states and eleventh out of forty African Union states evaluated.[513] The low overall ranking is worth evaluating from a few additional perspectives. South Africa has the highest income level of the four states evaluated, one of the highest levels of gross domestic product (GDP) per capita in Africa, and the highest total GDP in all of Africa. Investigating the ecosystem vitality and environmental health policies of South Africa may help explain this disparity between economic and environmental performance.

Air pollution levels that affect South African ecosystems are very low. South African air pollution policies ranked very high (90.4) individually and were comparable to those of its income peers (91.1) and geographic neighbors (89.6). These policies enhance the resiliency of South African and regional ecosystems as the impacts of climate change become manifest. However, the policies that South Africa has implemented towards water for ecosystems (41.7) are very suspect. South Africa's income (69.6) and geographic peers (58.8) all have substantially more effective policies with regard to water quality and water stress, and these deficiencies will increase South Africa's exposure and sensitivity to climate change impacts. Additionally, South African approaches to conserving biodiversity and habitats (44.8) are somewhat more effective than states with comparable incomes (38.9), yet they are not on a par with regional states (76.4). South African efforts to address climate change are also lacking individually (51.4) and in relation to states with equivalent income (68.6) and regional neighbors (77.2). In contrast, efforts to protect and enhance natural resources such as forestry, fisheries, and farmland (86.6) are succeeding both in relation to income (83.6) and geographic peers (76.4).[514]

South Africa's combined environmental health policy measure (81.8) is high in one respect and low in another respect. The geographic neighbors of South Africa garnered very low environmental health policy scores (43.0), while South Africa's income peers scored much higher (90.2). Clearly, the disparity within the region has many causal factors. Possibly, the differences between income peers indicate that South Africa has something to learn about environmental health from income peers that could also be shared with regional neighbors. In addition, South Africa belonged to a country cluster that was large geographically and divergent economically. This cluster also included states with high-carbon-intensive industries and electricity-generation challenges, but these states do show impressive environmental health efforts.[515] Overall, these low and relatively high scores for South Africa indicate dissimilar mitigation and adaptation efforts and disparate exposure and sensitivity levels to climate change impacts. The following in-depth analysis of the individual indicators exposes environmental sector strengths and weaknesses, illuminating specific South African vulnerabilities to climate change.[516]

South Africa has an impressively low environmental burden of disease, but the low overall burden of disease could be threatened by expansion of infectious diseases such as malaria, aided by climate change, into South Africa.[517] Efforts by South Africa to provide clean drinking water and sanitation are being addressed in some areas[518] but still need comprehensive improvements. In fact, in eastern and southern Africa, approximately 35 million people do not have access to improved water sources, and the largest proportions are in Mozambique, Angola, South Africa, Zambia, and Malawi.[519] Actions to reduce indoor air pollution and urban particulates are also showing success, and this will reduce the exposure and sensitivity to climate change impacts for individual South Africans. Additionally, South African policies on ecosystem air pollution are effectual. On the other hand, South Africa does face challenges managing water assets.[520] The water-quality policies for ecosystems are ineffective, and many ecosystems are suffering from water stress. Competing economic sectors, including industrial, energy, and agricultural, are placing severe pressures on renewable and nonrenewable water sources, and these pressures must be addressed if the resiliency of many unique South African ecosystems[521] is to be maintained in the face of increasing pressures from climate change.[522] Mitigation policies need to be improved if emissions are to be reduced in the energy and industrial sectors. If additional mitigation policies are not implemented, then adaptation policies must be designed to counter negative climate change impacts.[523]

Great Lakes Region (East Africa)

Kenya is evaluated from east Africa. Kenya's EPI score is 69.0, the same as South Africa, placing it twelfth out of forty-three African states and ninety-seventh out of all states. Kenya shares similar rankings with South Africa: tenth out of forty African Union states and seventh out of thirty-eight sub-Saharan states.[524] Nevertheless, Kenya has many striking economic and environmental differences that are, at times, polar opposites of South Africa. While South Africa is one of the more economically and technologically advanced states in Africa, Kenya is one of the poorest and least advanced. As a consequence, the challenges and opportunities to counter environmental vulnerabilities to climate change in Kenya are very distinct from South Africa's. These unique environmental vulnerabilities can be explored by examining the ecosystem vitality and environmental health of Kenya and then perused by analyzing the individual performance indicators.

The ecosystem vitality in some areas is very impressive and overall, is strong. The detrimental impact of air pollution on Kenyan ecosystems is practically nonexistent, and Kenya's individual score (99.7) is higher than those of its income peers (92.5), geographic neighbors (89.6), and the other three states evaluated. Additionally, Kenya's ecosystem water policy score (70.5) is also the highest of the four states examined and exceeds income (62.0) and regional peer (58.8) scores. This trend continues when biodiversity and habitat conservation efforts are examined. Kenya's ranking (89.0) is again the income (63.8), regional (62.3), and case-evaluation leader. However, the ranking leadership position is lost when Kenya's ability to protect and conserve natural resources is analyzed. Kenya (83.9) does rank better than income (78.1) and regional peers (76.4) but does not perform as well as South Africa (86.6). Last, Kenyan efforts to mitigate climate change (84.1) are also strong, besting geographic peers (77.2) but bested by income peers (85.5).[525] The strong performance in preserving and enhancing ecosystem vitality in Kenya helps reduce environmental vulnerability to climate change impacts by depressing exposure and sensitivity levels and by increasing the effectiveness of mitigation and adaptation capabilities. Kenya's stellar environmental performance, however, does not continue when the state's environmental health is examined.

The environmental health of Kenya is not a bright spot for state environmental leadership. While not as low as South Africa's (40.6), the 54.5 score does reveal some policy challenges and issues. Kenya does perform substantially better than income (43.3) and geographic peers (43.0), and this could indicate possible regional leadership opportunities.[526] However, the low economic performance of Kenya, similar to Nigeria, does indicate that some of these policy challenges and opportunities to reduce exposure/sensitivity and craft mitigation/adaptation policies will have stronger causal relationships to economic factors than to environmental factors. Also comparable to Nigeria, Kenya placed in a country cluster with developing states whose economies are in transition. Kenya, too, has poor environmental health scores, but because Kenya's economy is in transition, its carbon-intensity levels remain low.[527] Kenya's opposing ecosystem vitality performance and environmental health scores must be examined in more detail if the underlying forces that drive this disparity are to be revealed.

Looking inside Kenya's environmental policies at individual indicators reveals contrasts in policy effectiveness. Kenya has a significant environmental burden of disease, which adds to levels of exposure and sensitivity to climate change for all Kenyans. For example, malaria-carrying mosquitoes have recently been found in the highlands of Kenya for the very first time, and the spread has been tied to climate change.[528] In addition, water quality conditions in Kenya are poor because of sanitation and filtration challenges throughout the state. Widespread use in Kenya of wood and charcoal fuels for cooking pollutes indoor air, harms human health, degrades local environments, releases greenhouse gases, and devastates natural carbon sinks.[529] These core environmental health indicators explain why the health of Kenya's citizens is under severe and growing stress. However, the indicators of ecosystem vitality paint a more positive picture of environmental performance.

Kenya has made great strides protecting biodiversity and habitats. Activities to preserve critical habitats are strong and diverse and have significantly reduced risks to protected species. In addition, Kenya is working hard to diminish deforestation and burning of forests, grasslands, and croplands. Nevertheless, inefficient use of rainwater and biomass in combination with expansion of croplands into marginal areas has substantially increased land degradation in northwestern and south-central Kenya.[530] Kenyan efforts to enhance ecosystem vitality and mitigate climate change are augmented by its extremely low emissions per capita of greenhouse gases and low levels of industrial carbon intensity. In contrast, some sectors of enhancing ecosystem vitality need additional effort. Uncontaminated water is an essential ingredient in ecosystem health, but the growing competition for ever-scarcer water sources in Kenya decreases ecosystem resiliency and increases sensitivity to climate changes. More effective policies that will improve water quality and condense water stress are needed in Kenya if climate change exposure and sensitivity are to be lessened.[531]

CONCLUSIONS/RECOMMENDATIONS

The EPI rankings, on the surface, do not indicate the variety of paths to environmental performance individual states have taken. Specifically, "countries that have similar EPI scores may still have very different patterns across the 25 indicators and policy categories."[532] The detailed examination of underlying indicators did reveal succinct in-state variations and opportunities to assimilate possible best regional practices. In-depth analysis also exposed the lack of a close relationship between income-group-peer performance and individual state performance. Furthermore, EPI analysts performed cluster analysis that generated country clusters of states that had similar performance across the EPI measures. The country clusters could be used to identify successful environmental policy "best practice" models that are not geographically or economically based.[533] The evaluation of the four cases provided evidence of clear linkages between vulnerability to climate change and the environmental performance of each state. The ten policy categories overlapped the vulnerability measures in some sectors but also offered direct indicators of differing aspects of climate change–related policies. The model performed well and enhanced the analysis of the environmental vulnerabilities of each state to global climate change. Ultimately, the model enabled several state-specific conclusions and recommendations.

Overall, most Egyptian environmental health policies are helping reduce exposure and sensitivity to the impacts of climate change. However, more effort is needed to reduce urban particulates, increase access to adequate sanitation, improve water quality for ecosystems, and decrease water stress. In particular, water scarcity is the factor that most dramatically affects the resiliency and coping capacities of Egyptians to climate change.[534] Egyptian policies towards ecosystem vitality are more mixed. The greatest weaknesses within mitigation and adaptation policies are the efforts to reduce greenhouse emissions. More needs to be done to reduce the emissions from electricity generation, and Egyptian industries will have to decrease the carbon intensity of their activities if mitigation strategies are to succeed.[535] Egypt is a leader amongst geographic peers on efforts to reduce environmental vulnerability to climate change, but weak scientific, technological, and economic capabilities hinder Egyptian efforts to share policy design and implementation. International community assistance will be required to supplement Egyptian regional leadership in countering global climate change.

Nigeria has much to learn from peers on building resiliency to climate change impacts. The environmental health and ecosystem vitality in Nigeria is threatened by climate change. High exposure and sensitivity levels are evident in a lack of policies to protect carbon sinks (forests), and these high levels are compounded by weak mitigation and adaptation policies. Also, Nigeria's weak scientific, technological, economic, and development infrastructures make it difficult for new environmental policies that reduce the impacts of climate change to be designed and shared with regional neighbors. Furthermore, neither Nigeria's income nor geographic peers can offer much policy help towards combating climate change vulnerability, and outside assistance from the international community will be required.

The diversity of policies regionally indicate a leadership gap that South Africa could fill, and the leadership efforts could be based on South Africa's strong overall environmental health scores and economic strength. South Africa faces comparable water-management threats that Egypt faces, and climate models predict some parts of South Africa may experience significant losses of water runoff.[536] South Africa has excellent scientific and technological resources that could be used to develop, model, and implement environmental policies to reduce vulnerability to climate change. These policies could then be exported to regional neighbors for implementation and application.

The bipolar nature of Kenya's environmental performance reflects a contrasting vulnerability to the impacts of climate change. The bulk of the pressures from climate change will affect the health of Kenya's people. The health of Kenyans is being degraded by reduced environmental health, and as environmental health diminishes, exposure and sensitivity to the impacts of climate change increase. In contrast, Kenyans have taken considerable steps to improve the vitality of their natural ecosystems. These efforts will mitigate climate change impacts, augment adaptation activities, and reduce the severity of exposure and sensitivity to the consequences of global warming. If Kenya can receive assistance from the international community to improve environmental health performance and maintain policies that enhance ecosystem vitality, then Kenyans could effectively weather the climate change storm.

The research into the environmental vulnerabilities of these African states is incomplete without a complementary investigation into the social vulnerabilities of these states. In particular, the close relationship between poverty, corruption, governmental effectiveness, and environmental vulnerability to climate change needs further investigation. In sum, environmental vulnerability to global climate change is especially acute in Africa, and more research is required to help Africans formulate effective environmental policies that will reduce exposure and sensitivity levels and enable implementation of successful mitigation and adaptation policies.

CHAPTER 7

TERRORISM AND ENVIRONMENTAL SECURITY IN AFRICA

Kent Hughes Butts[537]

INTRODUCTION

In 1998 terrorists successfully attacked the United States embassies in Nairobi, Kenya, and Dar es Salaam, Tanzania, killing hundreds of people and calling attention to the spread of violent Muslim extremist ideology across the vast African continent. In the intervening years, many other terrorist attacks have been initiated in Africa by similar organizations, capitalizing on weak governance and inadequate security to promote their competing political ideologies. At the same time, the long-established pattern of rebellion, insurgency, and violent conflict against established governments has continued to play out across Africa. Quite often, these terrorist attacks and violent conflicts are rooted in unaddressed environmental issues and tensions over natural resources. This chapter seeks to clarify the relationship between environmental security and terrorism, highlights the importance of this relationship to those seeking to enhance institutional capacity and governmental legitimacy and prevent failed states, and offers recommendations on how the military element of power might be used to support civilian authority in addressing the underlying conditions that terrorists seek to exploit.

ENVIRONMENTAL SECURITY

Since the end of the Cold War, U.S. national security interests have been threatened primarily by regional instability. This instability is caused by a variety of issues, including socioeconomic disparities between urban and rural populations, religious differences, tribal and clan enmities of historic length, weak governments, and corruption. While environmental issues are generally not regarded as causes of conflict, they often serve as a multiplier effect, exacerbating existing tensions and fanning the flames of conflict. Thus, successful efforts to promote regional stability and prevent threats to U.S. national security interests need to be rooted in an understanding of how environmental issues relate to instability and, in particular, how they serve as underlying conditions that encourage extremist ideology.

Environmental issues become security issues when they affect U.S. national security interests. This may be either negative, a threat such as conflict over water or food riots, or it may be positive, using environmental issues as confidence measures between governments with territorial disputes. While some may argue that environmental security should not be characterized solely by its role in the state security processes, for the U.S. government and its agencies, it is the role of environmental issues in securing U.S. national security objectives that drives their behavior.[538] Thus, in this chapter, the interagency definition delineated by the U.S. Environmental Protection Agency (EPA) will underpin our discussions of the role of environmental security in combating extremist ideology and terrorism: "environmental security is a process whereby solutions to environmental problems contribute to national security objectives."[539]

This definition does not fail to recognize the importance of environmental issues at the global level, where the impact of humankind's behavior is threatening the security of the world. In fact, increasingly, these impacts are being felt on state and human security and are thus directly related to U.S. national security interests, as detailed in the *National Security Strategies* of the United States. It is the relationship of these effects to state security which underpins the *U.S. National Security Strategy* and allows national security strategists to include environmental security in their policy recommendations and plans. In his 1988 National Security Strategy address, President Ronald Reagan pointed out, "The dangerous depletion or contamination of natural endowments of some nation's—soil, forest, water, and air—will create potential threats to the peace and prosperity that are in our national interests as well as the interests of the affected nations."[540]

Environmental and natural resource issues are increasingly recognized as having played major roles in recent conflicts, fueling violent conflicts in eighteen regions over the last three decades.[541] According to the United Nations Environmental Programme (UNEP), "Attempts to control natural resources or grievances caused by inequitable wealth sharing or environmental degradation can contribute to the outbreak of violence."[542] Policy makers and strategists need to recognize the role of environmental issues in conflict and the creation of tensions. While it is rare to identify a violent conflict that was exclusively caused by environmental security issues, environmental security issues quite often exacerbate existing tensions and play a significant role in conflict.[543] The ongoing conflict in the Democratic Republic of Congo has been sustained by the interests of foreign powers and political actors in exploiting the country's significant reserves of diamonds, columbium, tantalum, gold, and other resources. However, environmental security relates to U.S. national security interests in ways other than its role as a threat.

Conflict over access to or control of natural resources that threatens regional stability may threaten U.S. national security interests. Yet, a more powerful role of environmental security may be as a confidence-building measure to promote bilateral or multilateral cooperation. Communication and cooperation between regional states with existing disputes may be created through a process of cooperation on common environmental problems. Working together to prevent transnational environmental crime, such as overfishing by foreign-flag trawlers or illegal logging, or preserving a common watershed may promote confidence building, lessen regional tensions, and enhance U.S. national security interests. In 1991 the United States and the Soviet Union cosponsored the Madrid Conference.[544] The objective of the conference was to develop a peace process between Arab countries and Israel. Because Israel refused to cooperate directly with the Palestinians, the conference created multilateral fora in which both Israel and the Palestinians could participate. Water and the environment were two of the five functional areas for multilateral cooperation. Water has also served as an area of confidence building between India and its two Muslim neighbors, Bangladesh and Pakistan, over the Ganges and Indus rivers, respectively.[545] In addition to preventing conflict, environmental issues may also serve as vehicles for peace building between former conflicting parties. The United States Pacific Command has used environmental security as an engagement vehicle to bring contentious states together, promote multilateral cooperation, and combat terrorism.[546] Their 2005 Seismic Disaster Preparedness conference brought India, Pakistan, and China to a high-level conference promoting regional cooperation in addressing this vital national security issue.[547]

The Westphalia system of the nation-state guided practitioners of foreign policy for most of the twentieth century. While it retains its importance and primacy among security strategists, the post–Cold War milieu has given rise to the concept of human security and an increased awareness of the impact of environmental issues on state legitimacy. The United Nations Development Programme *Human Development Report 1994* was a watershed in elevating the importance of the human condition to security studies.[548] Human security, rather than state power alone, increasingly determines the stability of the regions. The instability that is now the chief threat to U.S. national security interests increasingly comes from the inability of countries (often emerging democracies) to meet the demands placed on the political system by a population seeking freedom from want and freedom from fear. Governments that lack legitimacy are vulnerable to extremist ideology and more likely to fail. Environmental issues directly impact the ability of governments to provide for the needs of their people and are thus an element of human security and a variable in state legitimacy and tenure.

COMBATING TERRORISM AND EXTREMIST IDEOLOGY

Conceptualizing conflict as a continuum of phases running from pre- to postconflict has proven useful in identifying the processes necessary to promote regional stability and prevent failed states. Environmental security is an important variable in each phase of the continuum and contributes to our understanding of the role of environmental security in combating extremist ideology and terrorism. The conflicts in Afghanistan and Iraq have demonstrated the limitations of military power and the difficulty of successful regime change without establishing a government capable of providing for the human security needs of its population. Environmental security issues are now seen as an essential element of pre- and postconflict sustainable development and stability—so much so that the U.S. Department of Defense has crafted DOD 3000.05, *Military Support for Stability, Security, Transition, and Reconstruction Operations*, which states that the U.S. military forces will be capable of nation building as well as combat operations.[549] The U.S. Department of State has created the Office of the Coordinator for Reconstruction and Stabilization, an interagency organization dedicated to preventing conflict or building governmental capacity and creating stability in a postconflict environment; it focuses upon the capacity of the government to successfully address human security.[550] Stability is rarely attained unless the government accrues legitimacy from satisfying the demands placed on the political system by the people.

Army Chief of Staff General George W. Casey's White Paper defines the current and future strategic environment as being one of persistent conflict, fed by such variables as demographic changes, resource demand for energy water and food, urbanization, climate change and natural disasters, and failed or failing states.[551] He suggests that we must apply military forces to establish conditions in which security can be maintained, training, equipping, and employing indigenous security forces to enhance and empower local governance capabilities. The U.S. Army Field Manual 3-07, *Stability Operations*, defines essential stability tasks as protecting natural resources and the environment and supporting agricultural development programs by addressing water, food security, irrigation, health, energy, and diversity.[552] Other strategic Department of Defense documents have similar language.

The end of the Cold War removed superpower influence and guidance from many developing states. In the absence of military and economic support from the superpowers, institutions became ineffective, and the capacity of regional states to build and maintain legitimacy in the eyes of their people waned. As they struggled to meet growing demands placed on the political system, developing-country governments became more vulnerable to criticism from minority, dissident, separatist, and religious groups; this invited the rise of extremist ideology. The competition for scarce resources among religious, ethnic, and socioeconomically disenfranchised groups gave rise to internal conflict. States such as Afghanistan failed and were governed by groups with extremist ideologies and a willingness to use violence to promote their views and attack their enemies. As the *9/11 Commission Report* states, "When people lose hope, when societies break down, when countries fragment, the breeding grounds for terrorism are created."[553] Addressing the needs of the people is essential to preventing the spread of extremist ideology and terrorism.

Since the Bush administration's first *National Security Strategy* in 2002, the tools with which the United States may promote its national security interests have been defense, diplomacy, and development.[554] The Obama administration's secretary of state, Hillary Clinton, made it clear that the U.S. will continue this model in her confirmation-hearing testimony: "We will use all the elements of our power—diplomacy, development, and defense—to work with those in Afghanistan and Pakistan who want to root out Al-Qaeda, the Taliban, and other violent extremists...in the fight against terrorism."[555] Development, in particular, has been singled out by leaders of countries with large Muslim populations or an insurgent threat as critical in diminishing the underlying conditions that terrorists seek to exploit. Sustainable development is achieved where proper priority is given to its three underlying elements (economic, social, and environmental) and their interaction.

Malaysian Prime Minister Datuk Seri Abdullah Ahmad Badawi said in 2005, "I believe that we can address the problem of extremism and terrorism by delivering better and more widespread development."[556] He went on to point out that the Muslim world suffers disproportionately from a lack of physical and social infrastructure, high illiteracy rates, and lack of human development. He argued that a major factor in raising the dignity of the Muslim world population and increasing their voice globally is economic development. During the Bush administration, experts on combating terror often asserted that poverty, illiteracy, and lack of economic development were not causes of terrorism because the terrorists behind the 9/11 attacks were ideologically sophisticated, educated elites. Southeast Asian leaders disagree. Philippine President Gloria Macapagal Arroyo argues that a balanced approach to combating terrorism that involves direct military attacks and addressing the underlying conditions of terrorism is essential: "We have to fight poverty in the places where they can recruit their supporters."[557] The Republic of the Philippines' combating-terrorism strategy places great emphasis on cutting the roots of insurgency, which includes addressing multiple environmental security issues that give rise to popular protest against the government. As the *National Strategy for Combating Terrorism* states, "Many terrorist organizations that have little in common with the poor and destitute masses exploit these conditions to their advantage."[558]

Islamist terrorism is considered the primary terrorist threat to the West. It is easy to understand why intellectual leaders of the terrorist movement have success in portraying the Western developed countries as responsible for the poverty of the Muslim countries. The Muslim world is well behind the West in terms of development, and the West is easily blamed by terrorist leaders for phenomena such as climate change because of its high per capita consumption of resources.[559] Globalization and access to instantaneous media reinforce this perception. However, facts tend to support this argument and make it easy for those leaders to recruit the support from among the population they need to carry on their terrorist campaign. Among adult Arabs, 40 percent are illiterate; two-thirds of Arab women are illiterate. Thirty-three percent of the Middle East subsists on fewer than two dollars per day. The gross domestic product of Spain is greater than that of the twenty-two countries comprising the Arab League.[560]

Poverty may not, in and of itself, cause terrorism, but it certainly facilitates the operations of terrorist organizations, and in Africa, there is a strong link between environmental issues and poverty. As the U.S. *Counterinsurgency Guide* makes clear, "the basic wants, needs and grievances of the population may have little to do with the intellectual ideology of insurgent leaders, but may still be exploited to generate support."[561] In the wake of the 9/11 attacks, the United States drafted the *2003 National Strategy for Combating Terrorism*, which served as the framework upon which U.S. agencies produced their efforts to combat terrorism. The strategy had four elements: defeating terrorists and their organizations; denying sponsorship, support, and sanctuary to terrorists; defending U.S. citizens and interests at home and abroad; and diminishing the underlying conditions that terrorists seek to exploit.[562] Environmental security, particularly in Africa and the Middle East, has a significant role in combating terrorism because environmental security issues are often the underlying conditions that terrorists or insurgents seek to exploit.

TERRORISM AND INSURGENCY IN AFRICA

The pattern of development failure that has characterized Africa since World War II continues today.[563] For every bright spot like Malawi or Zambia, there are former bright spots like Cote d'Ivoire and Zimbabwe that are struggling to survive. The 6 percent African economic growth rate of 2007 is misleading; it was carried by a handful of oil- and mineral-producing countries benefiting disproportionately from high energy and commodity market prices and the trade largesse of China seeking equity ownership of mineral and energy concessions. Most of Africa struggles with failing health systems; a thin layer of professional personnel; widespread, vitality-sapping diseases like HIV/AIDS, tuberculosis,

and malaria; and an inadequate agricultural system to supply the bur-
geoning, dependent urban populations.[564] As governments struggle with
the twin evils of higher energy prices and corruption, extremist ideology
hovers, ready to take advantage of any failed or failing states. Extremist ideology, terrorism, and insurgency pose a threat to U.S. interests by
establishing training bases in ungoverned spaces, destabilizing countries
that are strategically important to U.S. national security interests, and
exploiting failed or failing states.

Most of the formal extremist Islamic groups are concentrated in northern Africa, where an increasingly dry climate, vast desert, and growing
youth bulge among the overwhelmingly Muslim population strain governmental capacities. These environmental conditions make food security
and access to freshwater beyond the reach of many regional governments.
Terrorist groups have long used such government shortcomings as a strategic communication message in undermining governmental legitimacy
and proffering their own extremist ideology.

Operating primarily in Algeria is al-Qa'ida in the Lands of the Islamic
Mahghreb (AQIM). This Sunni Muslim group was formed in 1998 as
an offshoot of the Algerian Armed Islamic Group under the name of the
Salafist Group for Preaching and Combat. Most of AQIM's targets are
Algerian government or security personnel, and the objective of the group
is to overthrow the Algerian government and create an Islamic caliph-
ate.[565] AQIM is also active in Mali and has launched attacks in Mauritania. Funding for the group comes from European sleeper cells and illegal
activities, and the group recruits new members across North Africa from
extremists in Libya, Mauritania, Nigeria, Morocco, and Tunisia. The intelligence community expects AQIM to broaden its areas of attack against
Western targets throughout the region and to return trained terrorists to
their home countries in North Africa and the Sahel.[566]

Somalia is a failed state in which the population struggles with lawlessness, drought, poverty, and unemployment. Environmental security
problems of Somalia have been recognized since the United States intervened in the early 1990s to address the food security crisis. Fundamentalist Islamic teaching offers a return to traditional values and a sense of

security and community that has great appeal in this environment. The Somalia Islamic Courts Council established control of southern Somalia in late 2006. Its military wing is the Harakat Shabaab al-Mujahidin al Shabaab. When the Ethiopian armed forces and Somali government drove the Islamic Courts Council from power in 2007, it resulted in a continued Ethiopian military presence within Somalia and a rationale for further violence. Al Shabaab has been at the forefront of a guerrilla war and insurgency against the Ethiopians and the transitional federal government as well as supporting aid organizations. Its terrorist attacks have targeted NGO aid workers, Ethiopia military, African Union peacekeepers, and members of the Somali government. Members of al Shabaab's leadership fought in Afghanistan with Al-Qaeda and are generally linked to the Al-Qaeda organization.[567]

Terrorism and violent insurgency are not isolated to Al-Qaeda–affiliated Muslim groups in the north of Africa. While the United States is focused primarily on a global Islamist insurgency, systematic terror has long been a staple of low-intensity conflict and insurgencies across the continent. Most often, these acts are motivated by domestic grievances, such as the Egyptian Muslim Brotherhood, which was driven by the belief that the principles of Islam were not being used to govern by Egypt's Muslim leaders. Socioeconomic conditions such as poverty, rising unemployment, illiteracy, and the concentration of wealth in an elite minority can give rise to hostility and bitterness and create "a breeding ground for alienation and radicalization."[568]

ENVIRONMENTAL SECURITY ISSUES IN AFRICA

Environmental issues do not cause terrorism; there are many regions of Africa with chronic or critical environmental security issues that do not have a terrorist problem. However, environmental issues often exacerbate existing tensions and are a critical variable in successful terrorist strategies. Quite often, these environmental issues relate to natural resources that have not been managed equitably or efficiently by the state government, leading to desertification, lack of clean water, disease, food riots, tensions between political and social groups or forced migration, and a loss of governmental legitimacy. Leaders of groups that wish to gain political power and erode the influence of a state government may portray these environmental problems as measures of governmental ineffectiveness and introduce a competing political ideology as a viable alternative to that of the host government. When a region or state is experiencing tension or conflict, environmental issues may multiply the problems and provide an opportunity for one or more groups to justify the use of violence or terrorism. Thus, it is important not to categorically state that poverty does not cause terrorism; it is misleading and provides unwarranted justification to those who wish to address the terrorist problem by focusing almost exclusively on the "kinetic" or attack-and-disrupt element of the combating-terrorism strategy and minimize the importance of sustainable development. It is essential to identify the underlying conditions in a given country or region and address those conditions that are robbing governments of legitimacy and may trigger violent conflict or motivate affected populations to support or tolerate ideological extremists. In the words of former Marine Forces in the Pacific Commander and Assistant Secretary for Asian and Pacific Security Affairs Lt. Gen. Wallace "Chip" Gregson Jr. (USMC, ret.), "By providing what local governments have not, these insurgents have gained legitimacy, psychologically conditioned these populations, and created an area from which they can safely operate."[569]

Food security is a major environmental security issue for much of Africa. Changing climatic conditions have brought devastating floods to destroy agricultural lands and seemingly endless droughts to promote desertification and the forced migration of thousands of Africans. Africa's vast geographic area makes food transport difficult and expensive, and this is complicated by the high levels of migration to urban areas, where subsistence agriculture is not possible for the large number of people living in squatter communities. The inability of governments to deal with these conditions and provide the basic nourishment that health requires is a fundamental measure of effectiveness that plays out repeatedly in political debates and elections as well as in the rise of extremist ideology. Ethiopia is a case in point. A landlocked state,

Ethiopia's food imports have to pass through the port of Djibouti, where congestion is stranding some 60,000 metric tons of food aid. The country is experiencing an acute drought that could make some 5.9 million people dependent upon donor food imports of nearly 600,000 tons.[570] Government opposition groups are highly critical of the government's inability to deal with poverty and provide food security, pointing out that Ethiopians are "victims of hunger and the skyrocketing food prices" and condemning governmental economic policies because "unemployment and hopelessness are rife among the youth."[571]

Other environmental security issues beset the full extent of the African continent. From Sudan to Nigeria, droughts and overgrazing have led to desertification and a migration to the south of Arab herders seeking grazing land. This migration brings them into areas of African farmers and often into violent conflict over access to the scarce resource of arable land; this has been a significant variable in the Darfur genocide. In Zimbabwe, the drought has led to widespread migration of Zimbabweans into South Africa, where they are faced with xenophobic violence. Nigeria's southern oil fields provide vital oil supplies to both China and the United States and significant foreign-exchange earnings to the Nigerian government. The unequal distribution of this oil wealth between the northern-dominated government and the poverty-stricken southern peoples has led to violent conflict and insurgent movements that have cost the government billions of dollars in lost petroleum revenue.[572] These north-south tensions have been fed by the environmental devastation, health impacts, and loss of agricultural land productivity created by the petroleum-recovery process.

Because of its broad impact, climate change is widely recognized as the most important environmental security issue. In the last two years, two major studies of climate change and security were conducted in the United States, the National Intelligence Council *National Intelligence Assessment on Climate Change and Security,* and the Center for Naval Analysis Corporation's *National Security and the Threat of Climate Change.*[573] Critical variables affected by climate change and examined in these studies included failed states, terrorist opportunities, social unrest and migration, unrest in Islamic countries, water and food security, health and disease, and stability of governments. The growing awareness of its importance to such variables has made climate change the top national security interest of Great Britain and Germany and a dominant concern for military and civilian government leaders from Africa to Asia.

Climate change affects the vital national security interests of many regional states. In northern and southern Africa, "drought is no longer a slow onset disaster but a chronic emergency."[574] The vast area of the Sahel is witnessing a shrinking of grasslands as warmer temperatures and reduced rainfall allow the encroachment of the desert. Subsistence agriculture underpins many African state economies and is directly affected by the increasing droughts and floods. Freshwater scarcity plagues northern, eastern, and southern Africa, and scientific predictions call for less rainfall in the future. Waterborne diseases such as malaria and dengue fever will spread with the new ecosystems created by climate change. More powerful storms and erosion will affect heavily populated coastal regions, and the resulting floods will have a significant impact on agricultural production.[575] Environmental refugees will continue their forced migration in search of water and food, often crossing borders and coming into conflict with other peoples, and demonstrating the weakness of African states. Climate change provides a striking example of how human security directly relates to state security and regional stability. When other necessary conditions exist, such as high population-growth rates, weak development, and a large Muslim youth bulge, climate change effects create conditions that are easily exploited by terrorist organizations. A common strategic communication message of Al-Qaeda is that climate change is the result of the rapacious consumption of scarce resources by the West, and the United States in particular, and that the impacts of climate change fall disproportionately upon the Muslim people.[576]

SUMMARY

The primary threat to U.S. national security interests is regional instability. Environmental security issues affect regional stability in three ways: they serve as a multiplier effect to existing regional or intrastate tensions; they may serve as confidence-building or peacemaking measures to help parties in ongoing conflicts such as territorial disputes to build communication and cooperation; and they negatively affect the human security of populations, posing threats to governmental legitimacy that lead to failed states. Environmental security issues affect every phase of the conflict continuum and are essential to achieving stability and sustainable development. As stated by the Environmental Protection Agency in its environmental security strategy, "Environmental security is a process whereby solutions to environmental problems contribute to national security objectives."[577] In Africa, a sample list of environmental security issues includes competition for resources, population growth, environmental degradation, food security, climate change, freshwater scarcity, loss of fisheries, and deforestation. Environmental security issues affect both state and human security concepts and often serve as a bridge between the two. In the case of combating terrorism, environmental issues often serve as underlying conditions that terrorists seek to exploit.

Addressing environment security issues in Africa should be an interagency effort that combines the resources of diplomacy, development, and the defense. The Department of Defense (DOD) has a rich history of addressing environmental security issues that runs from the role of the army and settling the West through counterinsurgency and nation-building efforts during the Cold War. Any U.S. effort aimed at combating terror on the African continent should be a coordinated effort that combines interagency capabilities to address not just known terrorists but the underlying conditions that terrorists seek to exploit. In Africa, those underlying conditions are increasingly environmental in nature, and this trend will only continue as the effects of climate change challenge governmental capacities. The capacity-building mission of the new U.S. Africa Command could be used to promote developmental assistance under the lead of USAID in regions where growing Muslim or disadvantaged populations face intense water scarcity, food security, and environmental degradation problems. Independent and mutually exclusive approaches to the kinetic and humanitarian dimensions of combating terrorism are an inefficient use of scarce resources and run the risk of creating a crisis of expectations and sending mixed strategic communication messages. In addition to helping U.S. forces run terrorists to ground, host nation militaries could easily develop the capacity to support their thinly staffed civil authority in addressing environmental security issues upon which governmental legitimacy depends.[578] Recognizing the environmental security dimension is essential to successfully dealing with terror and insurgency on the African continent.

PART III

RECOMMENDATIONS, REMEDIATION, AND CONCLUSIONS

CHAPTER 8

FROM GREED TO GRIEVANCE

NATURAL RESOURCES AND CONFLICT IN AFRICA

Elisabeth Feleke[579]

INTRODUCTION

According to the World Bank, Africa is home to only 10 percent of the world's population. However, roughly 30 percent of the world's poor are Africans.[580] Since the mid-1980s, the number of poor in Africa has doubled to some 300 million and is continuing to increase. Poverty is linked to environment in complex ways, particularly in those African economies that are heavily dependent on natural resources. About two-thirds of Africa's population lives in rural areas, deriving income from agriculture, mostly characterized by subsistence farming. As such, the condition of the local environment is a crucial determining factor in the ability of societies to survive. Michael Ross, in his analysis of the impact of natural resources on livelihoods, suggests that the greater a country's poverty, the more likely it is to face a civil war. He states that people are more likely to rise against their government when their economic predicament is deteriorating. Furthermore, rebel groups find it easier to recruit new members when there is widespread poverty and unemployment. It makes the prospect of combat and looting seem more attractive.[581]

Within the above context, the environment-conflict debate raises important questions: how might environmental factors directly or indirectly influence conflict or cooperation at the individual or community level, and how might environmental sustainability and security become a transnational issue in Africa? Although research on these issues is still lagging, there are several areas that have come into focus over the last ten years. A number of researchers have explored the "greed" versus "grievance" theories, more recently termed "resource abundance" and "resource scarcity" models of environment and security. While it is difficult to make a direct causal link between the environment and conflict, this chapter will argue that poor management and illegal exploitation of natural/environmental resources have a detrimental impact on security. More specifically, in the African context, both resource abundance and scarcity play a role in exacerbating and prolonging conflicts.

The resource abundance theory suggests that inequitable access to or unsustainable use of resources may lead to conflict. In Africa, the abundance of resources such as timber, diamonds, gold, coltan, and oil, which are portable and lucrative—often destined for Western markets—play a critical role in fostering environment-related conflicts. This is particularly true of countries in western and central Africa, such as the Liberia and the Democratic Republic of Congo (DRC). In turn, resource depletion or scarcity has been shown to play an important role in creating or exacerbating human insecurity, deepening ethnic divides, and straining governance and dispute resolution mechanisms. This is starkly evident in several countries in the Horn of Africa, such as Sudan and Somalia.

Resource Abundance and Conflict in Central and West Africa

Some studies of the resource-abundance phenomenon suggest that natural resource wealth tends to influence governments by making them less able or less willing to resolve conflicts and more likely to intensify them. This trend occurs through three mechanisms associated with state weakness in Africa. The first is corruption, where resource wealth often provides governments with easy and ready access to revenue. In many instances, this situation is compounded by a lack of or weak accountability, transparency, and oversight. In some African states, there is little or no capacity to effectively track large amounts of revenue generated through the extraction of resources. In such situations, corruption becomes rampant, and in fact, it is a common sentiment among many ordinary African citizens that some politicians run for office precisely for the possibility of personally amassing wealth, including that generated from control of natural resources. This is closely tied to the second mechanism. It is difficult for governments to provide law and order in the extractive regions since most valuable resources, such as gemstones and minerals, can be extracted without training or large investments. This opens the door for various nonstate actors to benefit from exploiting these resources, including some who eventually become powerful enough to challenge state authority. In addition, when governments raise their revenues through natural resources rather than taxes, they often fail to strengthen bureaucracies and institutions to support other aspects of their economies. The third and final mechanism relates to reduced accountability, where governments that rely on resource abundance are often less accountable and as a result, become less democratic. They often resort to using revenues to suppress opposition by dispensing patronage and building up their domestic security forces.[582] The best examples of all three issues discussed above, as well as the correlations between resource abundance and conflict, are found in the civil wars of the Democratic Republic of Congo (DRC) and Liberia.

The Democratic Republic of Congo

The DRC represents a good example where resources have played a role in (1) exacerbating and prolonging conflict, (2) worsening regional relations, (3) and seemingly contributing to the intractableness of the conflict which continues to ravage that country today. An estimated 3.3 million people died in the conflict between 1998 and 2004 in the DRC, mostly from war-related diseases and starvation.[583] The conflict saw widespread violence and massacres that United Nations peacekeepers were often unable to stop. Today, the impact of the exploitation of resources on security is clearly felt throughout the DRC, with rampant state corruption, weak institutions and bureaucracies, and reduced government accountability.

Two distinct patterns emerge when abundant natural resources exacerbate and prolong conflict. The first is the idea that natural resources are worth fighting over, and the second is the notion of natural resources as a source of funding for the conflict. Warfare is a costly business, and rebels in the DRC have long been successful in setting up financial administrative bodies in their controlled areas, especially with regards to trading with Rwanda and Uganda, while the government in Kinshasa has also been able to finance its side of the conflict from extractive resources. Whereas in previous decades, rebel groups largely relied on friendly governments to finance their activities, today, rebellion has become a private enterprise, where the market for natural resources allows these groups to become self-sufficient. This new stream of revenue permits various groups to purchase arms and recruit new members in order to continue fighting. The exploitation of natural resources has also led to worsening regional relations in the Great Lakes region. Researchers often refer to the conflict in the DRC as an international battle over natural resources. Due to the immense resource wealth of the country, various foreign powers as well as internal powers have sought to gain advantage. Countries such as Angola, Rwanda, Uganda, and Zimbabwe have all played a role in the conflict, and their involvement was sometimes driven and financed by the exploitation of natural resources. Arvind Ganesan and Alex Vines[584] state that the role of foreign governments who provide political, material, financial, or military support to various rebel groups and governments for their own economic interest is often overlooked but should be factored into the overall conflict analysis, prevention, and management.

To further elaborate, the Hutu interahamwe militia, which was largely responsible for the genocide in Rwanda, fled to neighboring DRC after a defeat by largely Tutsi-dominated Rwandan Patriotic Front forces. From there, the militia often launched attacks into Rwanda, prompting a Rwandan government invasion of the DRC. Rwanda justified its role in the four-year war as necessary for securing its border. Critics accused it of using the interahamwe attacks as an excuse to deploy 20,000 troops to take control of Congolese diamond mines and other mineral resources.[585]

A striking 2001 report by a UN panel of experts illustrates the example of coltan, a mineral used in mobile phones and computer chips. Given the substantial increase in the price of coltan between late 1999 and 2000, a period where worldwide demand was increasing, a kilo of coltan was estimated to cost $200. The Rwandan Defense Forces (RDF), through the Rwanda Metals Company, were exporting at least one hundred tons per month. The UN panel suggests that the RDF could have made $20 million per month simply by selling coltan, which translates into $250 million over an eighteen-month period. This was enough to finance and prolong the war and sustain the army's presence in DRC. The panel also identified Ugandan and Burundian rebels as parties to the looting and smuggling of coltan by using illegal monopolies and sometimes even forced labor.[586]

The same panel states that Uganda has benefited greatly from the DRC's gold and diamonds. Uganda has no known diamond reserves but became an exporter after it occupied diamond-rich areas in the DRC. Similarly, Uganda's gold exports increased dramatically during the conflict. The panel explains that in the Ituri region, rich with mineral reserves, Ugandan-backed insurgents played a direct role in the fight over control of resources. The Ugandan economy significantly benefited from the reexportation of gold, diamonds, coltan, timber, and coffee, which in turn led to an improvement of the country's balance of payments. However, in 2005, the International Court of Justice found the Ugandan People's Defense Force (UPDF) guilty of grave human rights abuses and the illegal exploitation of Congolese natural resources.[587]

Lastly, the exploitation of resources in the DRC has contributed to the intractableness of the conflict. Today, conflict remains in eastern Congo. Rebel groups, including dissident members of former rebel movements, Rwandan-backed Tutsi rebels, and Hutu militias with Democratic Forces for the Liberation of Rwanda (FDLR), continue to fight the government as well as rival groups, often seeking control of mineral wealth. As an example, in 2006, Uganda's rebel group, the Lords Resistance Army (LRA), settled near Garamba National Park in northeastern Congo. This move reignited tensions between Congo and Uganda as Kampala threatened to pursue the LRA into Congolese territory and Kinshasa suspected Uganda of wanting access to eastern Congo to exploit its natural resources.[588] Meanwhile, the human cost of this conflict continues to grow both in terms of the number of deaths as well as in the number of internally displaced persons (IDPs) and refugees.

Liberia
More than a quarter of a million people died, half of them civilians, and some 1.3 million people were displaced during the civil war in Liberia which raged from 1989 to 2003. A report from Global Witness states that the war in Liberia provides "a stark example of military-political entrepreneurship driven by natural resource exploitation."[589] Warlord Charles Taylor financed his armed uprising in 1989 by using revenue generated from the sale of timber and diamonds. When he gained power in 1997, Taylor proceeded to sponsor the brutal rebel movement, the Revolutionary United Front (RUF), in neighboring Sierra Leone. The same report concluded that the Liberian government provided material support to the RUF and sent Liberian soldiers to fight alongside them, partially in an effort to gain control of the lucrative Sierra Leonean diamond fields. These mines were less than one hundred miles from the Liberian border.

Two years after Liberia became involved in funding the war in Sierra Leone, the UN imposed sanctions on Liberian diamonds. As a result of these sanctions, Taylor's government shifted its focus to timber as a key source of funds. Taylor often bypassed normal state institutions, diverting logging revenues to himself rather than the treasury and using these funds to finance his ambitions. Although Liberia's revenues from logging were a minimum of US$187 million in 2000, according to government figures, only US$7 million were recorded. This meant that the government could not account for $100 million of revenues. The UN panel sent to investigate this discrepancy found that logging companies had hired militias who were acting as private armies. In addition, the Liberian timber industry played a vital role in arms brokering, with logging companies sometimes acting as arms traffickers.[590]

In 2000 Taylor passed the Strategic Commodities Act, which declared the president to have the "sole power to execute, negotiate and conclude all commercial contracts or agreements with any foreign or domestic investor." This act effectively gave him absolute control of all the natural resources in Liberia. This "theft by legislation" not only gave Taylor's government full authority to pillage but it also encouraged rapid liquidation of Liberia's natural resources, largely brought on by regime instability. For example, between 1997 and 2001, the production of roundwood in Liberia was estimated to have increased by more than 1,300 percent.[591]

In March 2003, the Special Court for Sierra Leone formally indicted Charles Taylor for participating in a criminal enterprise and for his efforts to gain access to the mineral wealth and to destabilize the government of Sierra Leone. UN timber sanctions were finally imposed in July 2003, more than two years after they were first discussed by the Security Council. With the majority of his funding cut off and with rebels rapidly advancing on Monrovia, Taylor fled to Nigeria and continued to influence politics in Liberia until he was arrested on March 29, 2006.[592]

The use of timber to finance conflict in Liberia will continue to have a negative impact long after the conflict has ended. In West Africa, Liberia was one of the few countries with a considerable amount of original rain forest cover. However, according to some experts, mostly as a result of the civil war and subsequent exploitation, the country has lost 50 percent of its forest cover and faces the likelihood of severe drought in fifteen years.[593]

The two examples discussed earlier illustrate clearly that rebel groups are not the only actors benefitting from the exploitation of abundant natural resources. Governments of resource-rich countries, often corrupt and unaccountable, grossly mismanage their economies. Revenues are often diverted for illegal purposes, and when conflicts do occur, these governments are often accused of grave human rights abuses. The international community has not been able to design a comprehensive strategy to address the problem of illegal exploitation of natural resources and its impact on security. Sanctions have not been adopted systematically and usually take a considerable amount of time to be successfully implemented. The need is therefore not only to address wars directly but also to address the trade that underlies them.

RESOURCE SCARCITY AND CONFLICT IN THE HORN OF AFRICA

Although most conflicts are brought about by a set of intertwined complex events, a number of studies suggest that natural resources increase the danger that civil war will break out and, more importantly, that it be will be more difficult to resolve. The correlation between conflict and resource scarcity is characterized, first, by environmental tensions such as the depletion and degradation of natural resources, including deforestation, overgrazing, water scarcity, and so on, and second, by conflict triggers such as population growth and movement, unequal distribution and poor management of resources, disease, and environmental disasters, among others. The Horn of Africa encompasses all of these elements. The region is experiencing degradation and depletion of resources at a rate faster than their renewal. The countries of the Horn have the highest population growth rates in Africa. In addition, countries such as Sudan and Somalia emerged in the 1990s from two decades of rule under pseudo-Marxist regimes with implications for environment and security. For example, nationalization gave the state monopoly over the production and distribution of material where the few available resources were concentrated in the hands of the few, leading to glaring disparities between classes, regions, and ethnic groups.[594] This, in turn, contributed to environment-related dynamics and stresses.

Sudan (Darfur)

Darfur is a barren, mountainous land just below the Sahara in western Sudan, and it is also currently described as one of the world's worst humanitarian disasters. Since 2003, according to the UN, fighting in Darfur has killed more than 200,000 people and made refugees of 2.5 million more. The conflict is typically characterized, mostly by Western media, as genocide, waged by the Arab Janjaweed and their backers in the Sudanese government against Darfur's black Africans.[595]

A number of studies have pointed out that the roots of the conflict may have more to do with ecology than ethnicity. Communities living on the poor and arid soil of the Sahel are caught up in an eternal fight for water, food, and shelter. The few pockets of good land have been the focus of intermittent conflict for decades between nomads (who tend to be Arabs) and settled farmers (who tend to be Arab and African.) According to the United Nations Environment Programme, rainfall has been declining in the Darfur region for the past half-century.[596] As the Sahara advances steadily south, converting soil into sand, the competition for scare resources will intensify and create fuel for prolonged conflicts.

The shifting dynamics of the fighting in Darfur illustrate why the analysis commonly used to describe the conflict as ethnic animosity between Arabs and blacks may be too simplistic, or at least, incomplete. Historically, because of Darfur's harsh, dry terrain, the region's Arab herders and its non-Arab farmers have had to work together to survive. The farmers allowed the herders' livestock on their land in exchange for goods such as milk and meat. As resources become more scarce—triggered in large part by severe and continuous drought in the 1980s—that history of cooperation became strained. This may have helped persuade some local Arabs and non-Arabs to join forces against the central government.[597]

To many, Darfur represents an opportunity for the international community to address a war that is largely fought over resources. Already, the fighting in western Sudan has spilled into Chad and the Central

African Republic. The massive refugee camps set up in Chad near the Sudanese border house concentrated populations that are too big for the land to support. The influx of refugees is exhausting water and firewood in local communities that depend on them for their own survival. One report relates an example where staff members from the International Medical Corps increasingly find themselves mediating conflicts between refugees and local farmers, who complain that the influx of refugees has ruined their land. Resources are simply insufficient to meet the over-whelming needs, and population movements are likely to provoke con-flict due to pressures on already fragile economic bases.[598]

In June 2007, a UN Environment Programme investigation concluded that peace and people's livelihoods in Darfur as well as in the rest of Sudan are inextricably linked to the environmental challenge. UNEP carried out the *Sudan Post-Conflict Assessment* at the request of the new Government of National Unity and the government of southern Sudan. The report concluded that "just as environmental degradation can con-tribute to the triggering and perpetuation of conflict, the sustainable management of natural resources can provide the basis for long-term stability, sustainable livelihoods, and development." It further states that the most serious concerns in Darfur remain land degradation, the spread of deserts southwards by an average of one hundred kilometers over the past four decades, and the overgrazing of fragile soils by a live-stock population that has "exploded" from close to 27 million animals to around 135 million.[599]

Somalia

The civil war in Somalia started as an armed conflict in 1991 and mainly as a result of the ousting of the repressive regime of Siad Barre. Although politics, the fractured nature of clans, and ethnicity play a role in the con-flict in Somalia, competition over access to scarce resources is perpetu-ating and prolonging the war, with devastating consequences. The lack of any recognized or strong authority to address these issues makes pros-pects for peace elusive. At the heart of the Somali conflict is the issue of land tenure. It is a basic resource driving clan economic activities. Some studies estimate that 59 percent of Somalis practice pastoralism, and 17 percent who occupy fertile regions engage in crop production. As such, the main natural resources in the country are livestock, cash crops, and charcoal. Marine resources and the potential of oil and mineral reserves, although undeveloped, represent viable resources.[600]

According to a World Bank report, access and control of land for crop cultivation, animal grazing, and use of water points are the drivers of the civil war. The report contends that

> one of the most chronic forms of conflicts occurs when pastoral communities clash with each other and with agricultural communities over grazing land and water points. Moreover, pastoralists challenge illegal enclosures that prevent them access to fertile grazing land, thus producing conflicts.[601]

Before the war, most clans relied on traditional mechanisms to resolve disputes. Traditional elders were often able to use customary laws and enforce them effectively. The civil war disrupted these traditional mechanisms and created massive demographic movements. Groups with access to arms forcibly occupied land held by generations of Somalis.

Somalia also suffers from severe environmental degradation and desertification. For example, livestock is considered a lucrative export commodity. However, as a result of an export ban, overstocking and over-grazing have become a common problem, which has created flashpoints of conflict. In addition, excessive harvesting of trees for commodities such as charcoal, an energy source, has led to deterioration in the environment and rapid desertification. This, in turn, reduces the available fertile land, a key requirement for a primarily pastoral-based economy. The problem of desertification is also compounded by a water crisis. Periods of drought in Somaliland and Puntland have caused large-scale migration, unemployment, and increasing poverty, leading to clashes between migrants and local inhabitants.[602]

Largely due to the lack of regulating mechanisms, clan groups lower or raise their level of identity in competition over resources. This is often used to mobilize their clan kin to acquire economic benefits necessary for survival or for increased power. They increasingly use force to settle disputes due to the lack of structures to regulate the use of common resources.[603] This phenomenon will continue to prolong the conflict in Somalia, and without due consideration, will undermine any peace process.

In conclusion, whether one looks at resource abundance or scarcity, conflict has multiple and complex impacts on development and on environmental well-being. Civil wars undercut or destroy environmental, physical, human, and social capital, diminishing opportunities for sustainable development. As a result, the increase in human vulnerability triggers a cycle of tensions and leads to massive migration of the affected population. In the end, unresolved disputes over access to resources for exploitation or for survival tend to weaken state capacity and legitimacy.

African Responses to Natural Resource Management, Conflict, and Cooperation

Africa is home to some of the largest deposits of natural resource wealth in the world. How this wealth can be managed to promote peace and sustainable development remains a complex and often elusive endeavor for policy makers. However, Africa also has a long history of regional cooperation with many subregional and regional institutions and frameworks. Most focus on political and economic issues, but a few do address the sharing and management of resources via multilateral environmental agreements, river basin commissions, and institutions working on biodiversity. One notable example of this cooperation is the Nile River Basin Initiative.

Ten countries share the basin of the Nile: Burundi, Egypt, Eritrea, Ethiopia, Kenya, Rwanda, Sudan, Tanzania, Uganda, and DRC. Approximately 160 million people depend on the Nile River for their livelihoods, and 300 million live within the ten basin countries. The population in this area is expected to double in the next twenty-five years. With the exception of Kenya and Egypt, all of the basin countries are among the world's poorest. Although Ethiopia is the largest supplier of the Nile upstream, Egypt and Sudan hold absolute rights to use 100 percent of the river's water under agreements reached between Egypt and Britain (the then–colonial power) in 1929, and between Egypt and Sudan in 1959. This means that Egypt must consent to other nations' use of the Nile's water. Egypt controls the region's most powerful military and fears that its upstream neighbors will reduce its supply by constructing dams without its consent. Other basin countries, especially Ethiopia, have contested the validity of the Egypt-Sudan treaty.[604] Former UN secretary general Boutros-Boutros Gali famously announced that "water wars" were most likely inevitable in the future.

However, rather than the prediction becoming a reality, in the mid-1990s, a hopeful trend started to emerge. Rather than conflict, there was greater recognition of interdependence and opportunities of shared management of water resources, and Nile basin countries agreed in principle that the situation must change. To reach a consensus, they developed the Nile Basin Initiative in 1999, with a mandate to achieve sustainable development through equitable and efficient use of resources. This initiative is unique in that it recognizes the importance of the role of other nongovernment stakeholders such as civil society in the discussion/negotiation process. To facilitate this process of inclusion, it created two mechanisms. The first is the civil society stakeholder initiative, and the second is the Nile Basin Discourse Forums. Both provide a venue to air out expectations and grievances. Most are hopeful that this model has the potential of creating needed dialogue and preventing hostilities from becoming conflicts.[605]

RECOMMENDATIONS

The international community can assist in creating an environment where resources are not illicitly exploited. The UN Security Council has imposed sanctions in the past, which proved somewhat effective in curbing resource exploitation to finance conflict. However, as major stakeholders in environment-related conflicts, African states must themselves ensure the viable and appropriate use of their own natural resources. To create long-lasting solutions, policy recommendations must demonstrate that states must have a stake and ownership of the decision-making process.

Link natural resource governance to security
States must ensure that natural resources are part of their national security strategy. By working with international multilateral institutions, they can ensure that their resources are properly mapped and recorded. The process not only assists in the economic development arena, it also encourages transparency and accountability by a variety of actors and can be used in negotiations with private entities.

**Strengthen regional and subregional mechanisms
such as the African Union, the Economic Community
of West African States, and Southern African
Development Community**
If strengthened, these organizations can address the illegal exploitation of resources by providing the regional mechanisms to assist in creating standards for resource governance and establishing a forum to share lessons learned and best practices. As these organizations continue to build their conflict-prevention and management mechanisms, they must also ensure and develop the capacity to monitor the exploitation and trade in natural resources.

Engaging peacekeepers is crucial
Engagement is especially important given the current stage of development of the African Union Standby Force. Countries should factor in the role of natural resources in conflict during potential peacekeeping missions. Peacekeepers should have a mandate to secure natural resources, especially if such resources have contributed to the conflict.

**Create environmental dialogue as a lifeline to peace
and postconflict societies**
Recently, it has been noted that when parties are involved in violent conflict over values and visions, environmental issues can be less divisive

and can provide practical means for cooperation and local development. Environmental issues may represent a platform for dialogue between warring parties. These are some of the arguments put forth by researchers such as Dr. Geoffrey Dabelko from the Woodrow Wilson Center. Environmental dialogue can help identify pathways through which competition over resources can be transformed into cooperation. A good case study of environmental dialogue and cooperation is the Virunga Conservation Area. The conservation area is comprised of the Virunga National Park in the Democratic Republic of Congo (DRC), the Mgahinga Gorilla National Park in Uganda, and the Volcanoes National Park in Rwanda. The area is significant both in terms of the broader ecosystem containing the Bwindi Impenetrable Forest as well as a habitat of the few remaining mountain gorillas in the world. However, the conservation area is also under severe pressure from ongoing civil unrest and an influx of a massive number of refugees, population density, poaching, deforestation, and so on. A report by the Woodrow Wilson Center recently stated that "the interdependence and indivisibility of human security, good governance, equitable development, and respect for human rights is clearly demonstrated in the Virunga Volcanoes case study."[606] The effort is bringing together government and nongovernment actors to conduct a consultative process with local communities in order to create mechanisms to address grievances related to natural resources management, to define and facilitate conservation initiatives, and to promote development, especially in the tourism arena, that directly benefits local communities. The effort has seen some success and has led to a number of state and international treaties. Behind this effort lies the idea of introducing environmental and natural resources management as a way of helping overcome political tensions by promoting interaction and technical cooperation. The case study illustrates awareness, not just among conservationists but also among security professionals and politicians, that "mutual dependence on transboundary resources provides the context for using these resources as a pathway to peacemaking."[607]

CHAPTER 9

Environmental Security in Peacekeeping Operations

Brent C. Bankus[608]

Throughout human history, people and countries have fought over natural resources. From livestock, watering holes and fertile land, to trade routes, fish stocks and spices, sugar, oil, gold and other precious commodities, war has too often been the means to secure possessions of scarce resources. Even today, the uninterrupted supply of fuel and minerals is a key element of geopolitical considerations. Things are easier at times of plenty when all can share in the abundance, even if to different degrees. But when resources are scarce—whether energy, water or arable land—our fragile ecosystems become strained, as do the coping mechanisms of groups and individuals. This can lead to a breakdown of established codes of conduct, and even outright conflict.

—Ban Ki-Moon, United Nations Secretary General, 2007

INTRODUCTION

In the past twenty years, the environment and its related areas of focus, such as global warming, climate change, human security, and natural resources, have received increased attention at the national, regional, and international levels. Beginning in 1994, the United Nations Development Programme's *Human Development Report* introduced the concept of human security, which suggests that global and regional stability increasingly turn on the welfare of the individual. Regional security is best achieved by providing "freedom from want and freedom from fear for the people, as outlined by the United Nations Declaration of Human Rights of 1948" for the people. Environmental security is one of the seven essential components of the human security equation, with the other six components being food security, economic security, health security, personal security, community security, and political security.[609]

The focus of this chapter is on environmental security and how it relates to regional security and subsequently, to peacekeeping operations. Additionally, the chapter describes a sampling of environmental security initiatives by the UN, the North Atlantic Treaty Organization (NATO), the European Union (EU), and the African Union (AU) at the strategic and operational levels, and a few operational/tactical-level initiatives that are or may become relevant to the security concerns of African nation-states in the future. The chapter also examines peacekeeping operations in Africa, analyzing how the environment and natural resources relate to the violence in several operations.

Before analyzing the role of environmental security in military operations, it is important to examine what notable leaders have said about the importance of the environment and natural resources in recent history. For example, Israel prime minister Ariel Sharon has been quoted as saying, "People generally regard 5 June 1967 as the day the Six Day War began...That is the official date. But, in reality, it started two-and-a-half years earlier, on the day Israel decided to act against the diversion of the Jordan [River]."[610] Israeli prime minister Yitzhak Rabin, also a great believer in natural resource importance, stated, "If we solve every other problem in the Middle East, but do not satisfactorily resolve the water problem, our region will explode."[611]

As well, the idea of natural resources and their possible effects on national security interests and conflicts goes further back than the afore-mentioned quotes. As the British navy was changing from coal to oil circa 1900, then–first lord of the admiralty Winston Churchill was quoted as saying, "Safety and certainty in oil lies in variety and variety alone."[612]

ENVIRONMENTAL SECURITY AND REGIONAL STABILITY AND SECURITY

The United Nations has continued to be concerned about links between the environment and conflict and has worked through several channels to reduce environmentally induced conflicts. According to United Nations Environment Programme (UNEP) research and field observations, the environment and natural resources contribute to the man-made disaster of conflict in three ways:

1. Conflicts occur over "high value" resources such as minerals, precious stones, fuel resources (oil and natural gas), and timber.
2. Conflicts occur over the direct use of resources such as water, forests, arable land, and wildlife.
3. Conflict may occur in countries whose economies are tied to a narrow set of commodities and which, in many cases, have a fragile central government.[613]

A study by the UNEP in 2009, entitled *From Conflict to Peacebuild-ing, the Role of Natural Resources and the Environment*, documented the importance of environmental-conflict linkages by noting that since 1990, no fewer than nineteen conflicts have been attributed to the exploitation of natural resources.[614] Further, over the past sixty years, no fewer than 40 percent of all intrastate conflicts can be linked to natural resources.[615]

In its June 3–14, 1992, conference in Rio de Janeiro (reaffirming the June 1972 Stockholm conference of the UN Conference on the Human

Environment and Development), the UN called on all member states to abide by a number of principles related to the environment. Of these twenty-six principles, Principle 24 addresses the environment during times of armed conflict and states, "Warfare is inherently destructive of sustainable development. States shall therefore respect international law providing protection for the environment in times of armed conflict and cooperate in its further development, as necessary."[616]

ACTIVITIES AND ACTIONS TAKEN BY NATIONAL, REGIONAL, AND INTERNATIONAL ORGANIZATIONS

Undoubtedly, one of the most active international organizations focused on environmental issues has been the United Nations. Established in 1972, the UNEP is the principle organization focused on the environment. Its mission is "to provide leadership and encourage partnership in caring for the environment by inspiring, informing, and enabling nations and peoples to improve their quality of life without compromising that of future generations."[617]

Within the UNEP, the Post-Conflict and Disaster Management Branch (PCDMB) is the responsible organization charged with identifying where the environment is impacted by conflicts and disasters, or where the environment is a factor contributing to conflicts and disaster impacts. Based in Geneva, Switzerland, with field offices in Kabul and Khartoum, the PCDMB has worked in postconflict settings such as Afghanistan, Sudan, Iraq, and Lebanon, in addition to natural disaster sites in Pakistan, Indonesia, Sri Lanka, and Maldives. Because environmental issues relate equally to natural or man-made disasters, environmental management and good governance practices are essential for long-term peace, stability, and security in any affected country or region, regardless of catastrophe origin.[618]

In addition to acting alone on strategic and operational environmental venues, the UN has participated in collaborative efforts to assist countries to manage environmental risks. The Environment and Security Initiative (ENVSEC) is a prime example. Begun in 2003, the initiative is governed by a management board from the five principal partner organizations, including the Organization for Security and Cooperation in Europe (OSCE), the UNEP, the United Nations Development Programme (UNDP), the United Nations Economic Commission for Europe (UNECE), the Regional Environment Center for Central and Eastern Europe (REC), and the North Atlantic Treaty Organization (NATO).

The purpose of ENVSEC is to assess environmental and security risks of a particular region or country, then to engage all stakeholders such as policy makers, environmental subject matter experts, and the private sector to develop projects to mitigate tensions and generate solutions.[619] As of the end of 2006, no fewer than fifty projects were initiated in the regions of eastern and southeastern Europe, the Caucasus, and central Asia, with a budget of $11.6 million.[620]

Besides strategic and operational initiatives, the UN has taken steps at the tactical level to ensure troop units are trained in respecting the environment of the host nation area by including directions in the UN Peacekeeping Operations' *Principles and Guidelines*. Chapter 9 of the document outlines policies and procedures to be followed by troop-contributing nations.[621] Also, since UN Security Council resolutions are including environmental security missions for the military force, such as in the DRC, the UN has initiated policies and guidelines in its new *Environmental Policy for UN Field Missions* and *Environmental Guidelines for UN Field Missions*. Both documents are in draft form and are currently due for final approval and distribution after May 2009.[622] Other nations as well, such as South Africa, are beginning to implement environmental security initiatives such as developing and using an environment-related doctrine and utilizing that doctrine during training exercises.[623]

Among regional organizations, NATO has a proven record on environmental initiatives and has been lending security assistance in Africa for several years, beginning in 1994 with the Mediterranean Dialogue with North African nations. Since many European countries remain in contact with their former colonies on the African continent, this NATO engagement policy is designed to advise and assist with challenges such as terrorism, drug trafficking, pirates, and immigration, among other topics. For example, NATO has provided logistical and airlift support to the African Union mission to the Darfur and supported the antipirate effort off the coast of Somalia.[624]

In addition to its recent assistance to the continent, NATO has long sponsored, conducted, or participated in a variety of venues, from conferences and seminars to commissioning studies on a variety of environment-related topics. For example, in July of 1985, NATO initiated a pilot study, *Aircraft Noise in a Modern Society*. Their 1987 landmark study on environmental awareness in the armed forces eventually led to the 1993 NATO *Environmental Policy Statement for the Armed Forces*. Most recently, on March 12, 2008, and as part of the Science for Peace Programme, NATO partnered with a number of other nations, international "think tanks," and organizations, including the OSCE and the UNEP, to sponsor a Security Science Forum on Environmental Security. The forum was divided into four sessions, and their related topics consisted of

1. global security concerns relating to environmental issues,
2. the role of international organizations in addressing the environment and security,
3. partner country priorities for the NATO SPS program, and
4. horizon-scanning exercise on environmental security issues.

Of the four sessions, session four had a particular focus on possible future environmental security issues and the link between defense and environmental security. Topics of discussion included environmental security and the defense analysis perspective, a future security environment study by NATO Allied Command transaction, integral monitoring as a tool for environmental security, and lastly, maritime security. From the topics discussed, it is clear NATO is making a concerted effort to address environmental issues at the strategic and operational levels and, most importantly, is addressing how the military can impact the environment of the host nation.[625] In addition, and as of this writing, NATO was exploring the idea of including environmental security–related doctrine at the operational and tactical levels.

In Africa, the most active international organization involved in peacekeeping is the African Union (AU). The African Union established the African Union Mission in Somalia (ANISOM). Due to the lawlessness in Somalia, particularly after the departure of the United Nations Operations in Somalia II (UNOSOM II) in March 1995, the AU took measures into its own hands and had approved through the United Nations Security Council UN Resolution 1744. This UN-backed resolution gave the AU the authority to establish a regional peacekeeping force in Somalia to establish peace and security and help rebuild their war-torn country.[626] However, ANISOM's mandate included more than efforts to law and order by AU peacekeeping troops maintaining law on the ground. Beginning in the 1990s, with local fisherman raiding foreign fishing trawlers off the coast of Somalia, piracy operations grew into a lucrative industry to procure funds to carry on hostilities by the various belligerent groups.[627] To stem this tide of illegal activity in and around the waters off Somalia, the AU is a leading proponent, in coordination with the Assembly of the International Maritime Organization and the UN, of urging the nation-states of the Comoros, Djibouti, Egypt, Eritrea, Ethiopia, France, Jordan, and a host of other regional stakeholders, to eradicate piracy from the high seas.

The United States has also sponsored a number of environmental initiatives over the past four decades that are relevant to Africa. One of the more important ones was the recent establishment of the U.S. Africa Command (AFRICOM), a four-star command established on February 6, 2007, with its headquarters in Stuttgart, Germany. The significance of this initiative has many facets, but by far, the recognition that peace and stability on the continent not only affects Africans but United States and international community interests tops the list.[628] The establishment of AFRICOM was the culmination of a multiyear process that presented a more comprehensive approach to Africa. Structured differently from the more established commands, AFRICOM represents a more robust representation from other U.S. government departments such as, but not limited to, the U.S. Department of State (DoS) and the U.S. Agency for International Development (USAID). AFRICOM was also designed to advise and assist African nations in building capacity within their own countries and regional organizations, such as the AU, to deal with regional security contingencies on their own.

The mission of AFRICOM is "in concert with other U.S. government agencies and international partners, [to conduct] sustained security engagement through military-to-military programs, military-sponsored activities, and other military operations as directed to promote a stable and secure African environment in support of U.S. foreign policy."[629]

There are currently several ongoing AFRICOM-sponsored programs focused on assisting African nations to promote regional stability. These programs include, but are not limited to, the Combined Joint Task Force-Horn of Africa; Operation Enduring Freedom-Trans Sahara; African Contingency Operations Training and Assistance (ACOTA); Africa Partnership Station (APS), Civil-Military Assistance, Health Programs; State Partnership Program; and the International Military Education and Training.

Of these programs, several are focused on environmental security topics. For example, from 1997 to 2001, through the Department of State, the U.S. ran the African Crisis Response Initiative (ACRI), with a focus on training friendly African nations in traditional peacekeeping operation missions such as convoy escorting, communications, and logistics. In 2001 the name was changed to African Contingency Operations Training and Assistance (ACOTA), and peace enforcement mission training was added to the venue. Since then, and due to changing requirements, training has expanded from traditional unit and staff training to include environmental security scenarios, such as inadvertently affecting the host nation water table with vehicle contaminants.[630]

Another U.S.-led program, the African Partnership Station (APS), was developed in response to requests by African nations for military-to-military or civilian-military maritime training. The main focus of the APS program is on environmental issues such as "infrastructure building and cross border cooperation to assist African nations in securing maritime regions and sovereign waters." However, APS also "addresses criminal activity, piracy, environmental and fisheries violations," as well as other human security concerns, such as health.[631]

ENVIRONMENTAL SECURITY AND AFRICAN
PEACEKEEPING OPERATIONS

For a variety of reasons, historically, Africa has been plagued by a series of inter- and intrastate conflicts. In their 2004 paper, Paul Collier and Anke Hoeffler, both of Oxford University, posited the problems of civil war on the continent were related to three economic factors relating to a nation's level of income, its rate of growth, and its structure. Essentially, if a country is poor, in economic decline, and is dependent on natural resource exports, then it faces an inordinate amount of risk of civil war.[632]

This proposition seems to be true for many African nation-states, since there are currently seven active UN peacekeeping missions there, the most of any continent. The missions include MONUC (Democratic Republic of the Congo), UNAMID (Darfur), UNMIS (Sudan), UNOCI (Cote d'Ivoire), UNMIL (Liberia), MINURSO (western Sahara), and MINURCAT (Central African Republic and Chad), of which the first three will be discussed.[633]

The resource-rich DRC, which has been a war torn area for decades, is a prime example of how natural resources contribute to regional instability as various state and non-state actors try to garner the profits from these resources to fund their causes can exacerbate is a prime example. The mineral-rich eastern region contains substantial amounts of gold, diamonds, and coltan (the combination of columbium and tantalum), and the lumber industry has been a prime focus of proxy militias of neighboring countries, sent to abscond the natural resources of the region. In addition to mining, the DRC also shows great potential in the areas of agriculture (with the main cash crops of coffee, palm oil, cotton, cocoa, rubber, and tobacco) and a robust inland fishing industry. The most promising areas for fisheries lay in the eastern Rift Valley lakes, rivers, and swamps of the Congo Basin.[634]

Unfortunately, due to an extended period of civil war, government corruption, mismanagement, lack of infrastructure, and general instability overall, the full potential of the DRC has yet to be realized. This fact has been verified through interviews with recently returning UN peacekeepers who were in the DRC from 2006–2007. Their mission of interdicting the exodus of natural resources has been difficult mainly due to corruption of government and military officials. Through official DRC government channels, agreements are brokered with neighboring Rwanda, which acts as a distribution point to other nations for "pirated" resources such as gold, diamonds, and especially, coltan.[635] Besides the obvious value of gold and diamonds, coltan is as valuable due to its widespread use in manufacturing everything from jet engines to computer chips.

Similar to the DRC, the Sudan also has long been beleaguered by civil strife, experiencing some fifty years of violence of varying degrees. Although since 2005, the Sudan has been ruled by a central government, the Government of National Unity located in Khartoum, with Omar Hassan Ahmed El Bashir as president, there is still abundant violence, particularly in the northern, western, and southern states of the western Darfur region.

The natural resources of oil and natural gas reserves and hardwoods can be counted as contributing factors to the conflict in the Sudan. However, long periods of drought coupled with mass migration, population growth, and access to arable land due to decertification are even more applicable in describing the reasons for instability. These and other environmental factors have led to competition and violence between agriculturalists, nomads, and pastoralists, all of whom depend on natural resources for their living.[636] The extreme violence in the Sudan and the Darfur region has led to at least 300,000 deaths, 1.6 million internally displaced persons, and at least 200,000 people who fled to Chad.[637]

Once a country promising to be an economic power in Africa, in September of 2002, the Cote d'Ivoire also fell victim to civil war. The conflict stalemated as the rebel group, Forces Nouvelles, held the northern part of the country and the government led by President Laurent Gbagbo controlled the south. Once again, the lure of access to natural resources and their revenues were contributing factors, as gold, cocoa, and cotton were the main focus.

While several peace attempts were tried, the latest signed in March 2007, the lucrative sale of commodities has left the country divided.

In point of fact, in 2005, the UN Panel of Experts published a report detailing how the Forces Nouvelles were using revenues from the sale of these natural resources to fund their cause. The government was no better, also using the revenues from the sale of cocoa to finance its war effort.[638]

CONCLUSIONS AND RECOMMENDATIONS

While not inclusive of either nation-states or regional or international organizations who have active environmental security programs, the chapter provides several insights. First, the environment, including global warming, climate change, and natural resources, is a contributing factor to regional stability, particularly in areas where the populace is tied to the environment for their livelihood. Second, nation-states and regional and international organizations are providing a variety of venues at the policy-making or strategic and operational levels for research and the implementation of programs of a nonkinetic nature to mitigate environment-related issues and tensions. Third, the military, in both traditional and stability or peacekeeping operations, has proven to be a valuable asset. While the primary reason for military units is to establish and maintain security, they also bring a plethora of other assets, such as an established command structure, equipment, manpower, and expertise, that are invaluable in assisting civil authorities during natural or man-made calamities.

The importance of engagement among all stakeholders on environmental topics and regional security through national, regional, and international organizations, the private sector, and the military in at-risk areas cannot be overstated. Disasters, both natural and man-made, are not new phenomena and are not likely to decrease. It is therefore suggested that national, regional, and international organizations and the military should continue collaboration through a common working group, panel, or Red Team focused on environmental issues and should establish a central location through a database. The database should outline these initiatives and be managed by an oversight committee of mutual agreement. At the operational and tactical levels, it is recommended that all national armies, particularly troop-contributing nations of the UN, NATO, the European Union, and the AU, evaluate lessons learned from previous operations and determine what, if any, environmental missions should be added to their Mission Essential Task Lists. Further cooperation among the military and NGO, IO, and other private organizations should also be an area of focus. To that end, training sessions outlining what each organization does and is responsible for are invaluable. For example, the United States Agency for International Development (USAID) periodically conducts a Joint Humanitarian Operations Course (JHOC) in Washington, DC. The JHOC is designed with the joint environment in mind and focused towards training military personnel interacting on the modern battlefield and with the various nonmilitary entities they are likely to encounter. Since the USAID JHOC is only conducted periodically and there are no follow-up courses, it is recommended the JHOC be conducted more frequently and an advanced JHOC be developed and implemented.

CHAPTER 10

THE KAVANGO-ZAMBEZI (KAZA) TRANSFRONTIER CONSERVATION AREA PROJECT

Dan Henk [639]

INTRODUCTION

One of the most ambitious, interesting, and potentially significant political initiatives in contemporary Africa is the Kavango-Zambezi (KAZA) Transfrontier Conservation Area, a five-nation project formally initiated in 2004 that still was largely in planning in 2009. The scheme proposes to tie the national parks and wildlife conservancies of Angola, Botswana, Namibia, Zambia, and Zimbabwe into a common conservation area. The initiative consists partly of an enormous human development scheme, partly of a commitment to protect fragile ecosystems across a vast land area, and wholly of an extraordinary vision of a shared economic future.

By 2009, it had attracted the attention of foreign partners, businesses, scholars, and advocacy groups, along with the inevitable critics.

In its most optimistic projections, KAZA could greatly boost the economies of several African countries and help rehabilitate Angola's devastated interior. It is not difficult to imagine that its effects might ultimately expand into the war-ravaged Democratic Republic of the Congo, affording opportunities for desperately needed economic development and infrastructural rehabilitation in Africa's devastated heartland. KAZA proffered tantalizing glimpses of a significant new mechanism for regional integration, resource sharing, and conflict avoidance.

However, in 2009 the project still was far more a dream than a settled reality, with many potential pitfalls and unanswered questions. The complexity of the economic integration it envisioned was matched by few other initiatives in southern Africa. Despite the appearance of less ambitious regional precedents such as the Southern Africa (electrical) Power Pool, it was by no means certain that KAZA's participating countries ultimately would be willing to surrender so many of their economic prerogatives to a more centralized regional control or look favorably upon the economic empowerment of local communities. To facilitate tourism, the project envisioned a much freer flow of traffic along the regional transportation infrastructure, diminishing state control and remuneration. Such measures threatened the equities of some government elites. Nor were all environmentalists happy with the commercial complexion of the biodiversity objectives. The coordination of infrastructure, human development, and wildlife conservation called for considerable planning sophistication and high levels of managerial proficiency. It was unclear if southern Africa could muster the requisite expertise to implement such a complex and multifaceted plan.

Then, there was the troubling issue of security. In a region still awash in the detritus of the liberation wars of the late twentieth century and the civil wars of the early twenty-first, armed criminals, poachers, and possibly even terrorists stood to gain from open borders, spotty surveillance, and unrestricted movement across southern Africa, a movement facilitated by the generally good regional transportation infrastructure.

FIGURE 10.1. Southern African Development Community (SADC) Transfrontier Conservation Areas, current and proposed.

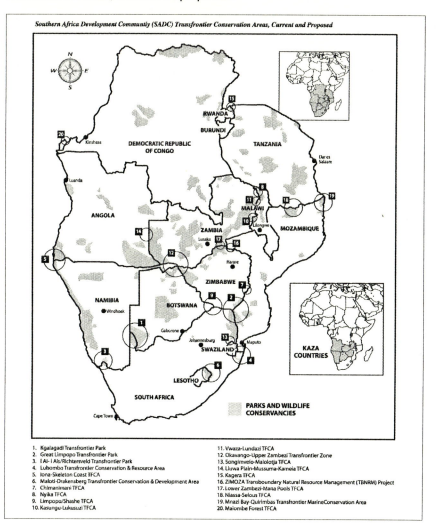

Southern Africa Development Communtiy (SADC) Transfrontier Conservation Areas, Current and Proposed

1. Kgalagadi Transfrontier Park
2. Great Limpopo Transfrontier Park
3. I Ai- I Ais/Richtersveld Transfrontier Park
4. Lubombo Transfrontier Conservation & Resource Area
5. Iona-Skeleton Coast TFCA
6. Maloti-Drakensberg Transfrontier Conservation & Development Area
7. Chimanimani TFCA
8. Nyika TFCA
9. Limpopo/Shashe TFCA
10. Kasungu-Lukusuzi TFCA

11. Vwaza-Lundazi TFCA
12. Okavango-Upper Zambezi Transfrontier Zone
13. Songimvelo-Malolotja TFCA
14. Liuwa Plain-Mussuma-Kameia TFCA
15. Kagera TFCA
16. ZIMOZA Transboundary Natural Resource Management (TBNRM) Project
17. Lower Zambezi-Mana Pools TFCA
18. Niassa-Selous TFCA
19. Mnazi Bay-Quirimbas Transfrontier MarineConservation Area
20. Maiombe Forest TFCA

Source. Drawn for the author by Jack Durham.

These circumstances seemed to require a robust regional security agenda, but in 2009, KAZA's planners were placing little priority on identifying and resolving the project's likely security-related dilemmas.

The limitations in KAZA security planning were a logical reflection of its key participants. The guiding hand in the whole effort was an environmentalist advocacy group, a nongovernmental organization based in South Africa. The most important planning roles for the project itself had accrued to the environmental ministries of the five participating countries. Conspicuous for their almost-total absence were the military, police, and intelligence agencies, an exclusion that was not inadvertent. The presence of the security agencies would have incurred a risk of "securitizing" the planning by holding the developmental aspects hostage to narrowly defined concepts of state sovereignty and border security. But the absence of security-sector planners nonetheless almost guaranteed that important questions were not being fully considered.

Any comprehensive listing of all the open questions surrounding KAZA in 2009 is a task well beyond the scope of this chapter. However, southern Africans and their partners around the world clearly stood to gain by a KAZA success, and the issue of security was the "elephant in the living room" that could not be permanently avoided. The real dilemma was how to get just the right amount of and just the right kind of security. This chapter describes the KAZA initiative and calls attention to the potential contribution that security-sector planners could make. Ultimately, it recommends their inclusion and identifies a role for external partners that could assist southern Africans in resolving KAZA's inevitable security problems.

BACKGROUND

The KAZA project did not appear out of whole cloth. It evolved from a variety of southern African initiatives with origins in the late twentieth century. To understand KAZA, it is necessary to have some appreciation for its antecedents.

Since the early 1990s, Africa has been cited by journalists and scholars as a stereotypical locale for environmental catastrophe, provoking Malthusian predictions of imminent social trauma and humanitarian crisis.[640] Few knowledgeable people doubt that Africans face severe environmentally related problems.[641] However, southern African countries have undertaken efforts to attenuate these problems. Botswana, Namibia, and South Africa each have surprisingly visible, politically influential environmentalist constituencies. All of these countries are committed by public doctrine to wise stewardship of the environment.[642] Since the 1990s, they have cooperated with one other and with other southern African nations in a variety of environmental initiatives encouraged by the Southern African Development Community.[643]

In the early years of the twenty-first century, southern Africa still faced daunting challenges. Biodiversity was under threat from urbanization and population growth. Regional megafauna (particularly elephants and rhinos) were vulnerable to commercial poaching. Pollution of water, air, and soils was rampant. Water continued to be a scarce resource with a potential to provoke violent competition. One of the most intractable dilemmas was the inherent tension between development and conservation. Countries in the region had limited resources for environmental protection, and populations still struggling to achieve the basic material necessities of life simply would not place an exclusive priority on the natural environment. Appeals to either a general concern for healthful conditions or an emphasis on the importance of biodiversity simply did not resonate with the majority of Africans. An environmental agenda appeared to have little prospect for success unless it offered direct and unambiguous near-term material incentives. Yet, despite these daunting challenges, southern Africa had seen tangible success in environmental initiatives by the turn of the century. When the UN organized its second worldwide environmental conference (in 2002), it chose Johannesburg, South Africa, as the venue, partly in recognition of regional environmental achievements.[644]

An interesting feature of the 2002 Johannesburg conference was the contention between two competing environmental visions, a difference that also marked something of an ideological divide between the developed and the developing world. At issue was the centrality of man to the environment. On one end of this spectrum were those activists whose "environmental" perspectives emphasized biodiversity. They downplayed the primacy of human beings and reflected what might be characterized as an ecocentric approach to environmental issues. On the other end of the spectrum were those who tended to see human beings—and their welfare—as the central feature of environmental issues, reflecting what might be characterized as an anthropocentric approach.[645] Environmentalists in Africa tended to the end of the spectrum emphasizing the centrality of mankind and were inclined to favor wise stewardship of the environment to meet human needs.

The ideological differences evident in UN-sponsored conferences are important to biodiversity in Africa. In the early years of the twenty-first century, the traumas and humiliations of European colonialism still colored African sensitivities. Africans recalled the arrogation by colonial authorities of almost limitless prerogatives to exploit human and natural resources. Africans remembered colonial regimes notorious for predatory exploitation of natural resources and equally infamous for abusive treatment of indigenous people. Those European colonial authorities interested in conservation often were dismissive of indigenous priorities and values, resulting in the establishment of parks, wildlife conservancies, and game preserves by arbitrary (sometimes coercive) eviction of African inhabitants from ancestral lands or by outright denial of traditional land use rights. Perhaps even more unfortunately, the indigenous authorities that rose to power in postcolonial Africa often copied the conservation practices of their colonial predecessors.[646]

CONTROVERSIES OVER "PROTECTIONISM"

By the late twentieth century, a loose, worldwide coalition of scholars and activists had begun to publicize the human cost of such policies. They challenged a peculiarly "European" concept of wildlife conservation that they argued was characteristic of the colonial regimes and was still being applied unfairly and coercively in Africa. They viewed this kind of conservation as an insidious and persistent form of colonialism. These scholars disputed the notion that either colonial authorities—or their successors in the governments of independent states—should have the prerogative to arbitrarily designate regions of a country for wildlife, abrogating the traditional access of local people to the land's resources.[647]

The scholars saw a continuing travesty. Early villains were colonial governments that alienated peasant land to reconstruct a peculiarly European vision of a "natural" Africa void of African people. Modern villains were African governments (and supporting environmental advocacy groups from the developed world) that sought the same ends. The scholars were especially contemptuous of what they called "protectionist" conservation programs—the use of violence by state authorities to deter "unauthorized" use of wildlife and other resources from the land.[648] This new thinking had a political impact in Africa and elsewhere. One outcome was increasing government sensitivity to accusations of "protectionism" in environmental programs. Another was an interest in (local) community-based natural resources management projects intended to reconcile national conservation interests with local land use rights.

COMMUNITY-BASED NATURAL RESOURCE MANAGEMENT

The Community-Based Natural Resource Management (CBNRM) programs were a significant regional environmental initiative.[649] They burst onto the southern African scene in the mid-1980s, heavily sponsored by the U.S. Agency for International Development (USAID) and other foreign donors. The programs sought to empower local rural populations by allowing them to manage the natural resources of the land on which they lived, providing them the right to funnel the economic benefits back into the local community. The U.S. government was an early subscriber to these programs and offered the most notable financial contribution to the ones in southern Africa—a multimillion-dollar effort conceived and managed by the regional USAID office, then based in Harare, Zimbabwe.[650]

Between the late 1980s and early 1990s, CBNRM programs were launched in Zimbabwe, Zambia, Botswana, and Namibia (four of the five KAZA countries). Though similar in intention and concept, each program differed in local detail and ultimate achievement. Despite the enthusiasm with which they were greeted (and the extensive support of foreign donors), the CBNRM programs in southern Africa soon encountered significant obstacles.[651] Zimbabwe's program (called "Campfire") produced several years of gratifying community mobilization and economic development but ultimately fell afoul of a national government that feared the political and economic empowerment of local communities at its own expense.[652] Zambia's program, called ADMADE (Administrative Management Design for Game Management Areas), likewise made an initial splash but appeared to be somewhat dormant by the early years of the twenty-first century.[653] Botswana's program still was functioning but was limited almost exclusively to one small area of the country and had incurred increasing government skepticism about its viability.[654] In southern Africa as a whole, only Namibia had a national CBNRM program with all the earmarks of an unqualified success.[655]

While the CBNRM programs were not spectacularly successful in southern Africa, they did reflect a concern by scholars, activists, and some government leaders for connecting human development in local rural communities to the exploitation of local natural resources. Whether or not this concern was evident in practice, it was thoroughly embedded in the official policy documents of southern African countries, so it enjoyed at least the status of an espoused value.[656]

WATER ISSUES AND OKAVANGO

Another important environmental issue in southern Africa has been access to water. Large parts of the region are desert or semidesert, and this already dry area faces a long-term trend of decreasing rainfall.[657] Meanwhile, demand for water is growing throughout southern Africa, posed by mounting human populations, expanding industry, and increasing agricultural demand. Rivers are key local sources of water, and they are almost by definition natural resources with security implications.[658]

Despite the almost inevitable potential for competition and conflict over this vital resource, water sharing is one of southern Africa's success stories, illustrated in the three-nation Permanent Okavango River Basin Water Commission, better known by its acronym, OKACOM.[659]

Controversy over water resources erupted between Botswana and Namibia in the early 1990s, immediately after Namibia attained its independence from South Africa. The main issue was the use of water from the Okavango River, a watercourse that drains a large area of south-central Angola before crossing Namibia's Caprivi Strip and ultimately emptying into Botswana's Okavango Delta. Okavango water is critically important to the ecology of the delta (and, by extension, both to the wildlife and the ecotourist industry in Botswana that exploits it). However, the same water also is a vital resource for Angola, eager for human development in its war-ravaged interior. It is of equal concern to Namibia, which covets any available water for its very dry capital region. Diversion of Okavango water by Angola or Namibia would have an enormous impact on Botswana and conceivably could fuel future conflict. However, the national leaders of both Botswana and Namibia clearly foresaw this danger in 1990 and established a Joint Permanent Technical Commission to deal with bilateral water issues. In 1994 Angola joined the group.

The extensive (and growing) regional consultation over the use of Okavango water now seems more likely to produce cooperation than conflict. In fact, this is an interesting African story about mutual consultation, conflict avoidance, and coordinated human development.[660] OKACOM did not solve all of the regional issues.[661] Access to water has been a contentious interstate problem in southern Africa, a region with a number of unresolved disputes, but OKACOM illustrates a southern African capacity to achieve negotiated solutions to such problems.

TRANSFRONTIER CONSERVATION AREAS

Other regional environmental initiatives appeared in southern Africa after South Africa began its transition to majority rule in 1989. One involved the work of a South African businessman, environmentalist, and philanthropist, Anton Rupert, who founded the Peace Parks Foundation in 1997, with its headquarters and staff offices in Stellenbosch.[662] The organization quickly became a savvy environmental advocacy group, with close government ties and an impressive capacity to attract international environmental celebrities, including Prince Bernhard of the Netherlands and South Africa's own iconic leader, Nelson Mandela. The centerpieces of the organization's activities have been transfrontier wildlife conservation areas, commonly known by the acronym TFCAs.[663]

TFCAs began to appear in southern Africa in the late 1990s, spurred by the Peace Parks Foundation (and a growing list of partners). The first half-dozen or so were established along South Africa's borders. They started as bilateral or trilateral programs in which southern African countries agreed to jointly manage national parks and wildlife reserves along contiguous frontiers. The projects produced immediate results: they were a boon to wildlife management and served laudable biodiversity ends. They also attracted environmentalists and brought international attention to southern African conservation. From the beginning, their advocates saw them as mechanisms for promoting regional cooperation and human development, reducing the prospects for interstate conflict, and contributing directly to the economic development of local communities that would benefit from tourism and resource harvesting.[664]

Perhaps not unexpectedly, the concept generated a fair amount of criticism. Some scholars argued that the initiatives were the obsession of small coteries of ecocentric government elites and international environmentalist groups whose desire for pristine wildlife refuges undermined any real commitment to local community development. Others found the principles of human security violated by government "top-down" approaches that failed to involve local communities in defining their own future. Still others argued that the promise of regional cooperation fell far short of the more optimistic expectations.[665] While these criticisms had merit, the reality fell somewhere between the dreams of the advocates and the disparagements of the detractors.

The Appearance of KAZA

KAZA is the most ambitious regional scheme to have sprung from the TFCA movement, and it is qualitatively different from its predecessors. Until the advent of KAZA, southern African projects had connected the contiguous areas of two or three countries. The most complex of these, the Great Limpopo Transfrontier Park, christened in late 2002, linked the corners of Mozambique, South Africa, and Zimbabwe. This park encompassed some 35,000 square kilometers of land between the three countries.[666] However, even the sprawling Great Limpopo Park is dwarfed by the KAZA project, with its 278,000 square kilometers of animal-rich savanna, woodlands, rivers, and wetlands (a land area similar in size to the nation of Italy) spread out in separate locations across five southern African countries. This new project included some thirty-six national parks, game reserves, wildlife management areas, and community-based conservancies, along with the Okavango Delta (the world's largest Ramsar site) and Victoria Falls, one of the seven natural wonders of the world.[667] And while KAZA is conceptually related to the other transfrontier conservation areas, earlier TFCAs were just one source for its inspiration. The project gathered many of the threads introduced in previous environmental initiatives, including the CBNRM emphasis on local human and economic development and the OKACOM emphasis on multilateral natural resources management. It also drew insights and advice from a worldwide environmentalist community.[668]

KAZA began to take its current form in 2003 and was formally launched by an assembly of the environmental ministers of the five participating countries in the Namibian town of Katima Mulilo in July 2004. The ministers agreed initially that the scheme would be overseen by an interim secretariat (located in Harare, Zimbabwe), with responsibility for details delegated to a joint technical management committee. Each of the five contributing countries was then tasked to pursue a separate aspect of the initiative.[669]

KAZA's initial articulated objectives included the joint management of regional natural resources, harmonized land use (with scientific monitoring and research), rationalized policy and legal frameworks, promotion of sustainable tourism, encouragement of public- and private-sector investment, and joint marketing. The ultimate goal of these efforts was economic development and a world-class tourist destination.[670]

By mid-2006, a draft memorandum of understanding (MOU) for KAZA was circulating amongst the five countries. In June 2006, representatives gathered in Luanda, Angola, to assess the status of the initiative. At that time, the Peace Parks Foundation offered a feasibility study drafted by Sedia Modise, the Peace Parks representative in Botswana. The study was well received, and the environmental ministers of the KAZA countries signed a final version of the memorandum at Victoria Falls, Zimbabwe, on December 7, 2007.[671] The result was a spate of follow-up meetings to work out the implementing details, a process that was still occurring as this chapter was written in early 2009. However, insiders close to the various committee deliberations told the author that progress had been slowed by political turmoil in Zimbabwe (as various key KAZA actors were waiting to see how the political dynamics in that country would play out).

By 2009, OKACOM and the earlier southern African transfrontier conservation areas had a track record that could be evaluated. However, the KAZA initiative still was largely a future dream, one considerably more ambitious than anything previously attempted. It connected some very dissimilar political actors, including countries with wide variations in economic development and overall governing coherence. The national government of one of the five countries, Zimbabwe, seemed bent on self-destruction, and Zimbabwe's presence as one of the KAZA signatories posed problems for potential external contributors like the United States. In fact, the KAZA initiative was so new that the various economic, infrastructural, and managerial relationships still required definition.

Then, too, the primary motivation behind the initiative was bound to be troubling to some of its potential supporters. Human and economic development was the primary articulated objective of KAZA. This orientation bothered some environmentalists, who feared "commercialization" of the environment and harm to fragile African ecosystems exploited for tourist dollars. Even so, by tying them to direct material benefits for local populations, the initiatives probably represented the last, best hope for safeguarding biodiversity in southern Africa. Whatever the reservations, by 2009, TFCAs in southern Africa had mobilized a very wide range of supporters, attracting coalitions of unlikely partners, ranging from policy makers to environmental scientists, development-oriented NGOs, and conservation advocacy groups.[672]

One feature of KAZA should particularly be highlighted. This has been a southern African initiative, conceived by southern Africans and locally implemented. Whatever inspiration and support it receives from external sources, it nonetheless reflects the case of a region seeking control of its own future. The high degree of local initiative distinguishes KAZA from most other large-scale environmental schemes on the continent, such as the Congo River Basin Initiative (whose supporters and advocates were claiming as a major environmental success by 2007).[673] One troubling aspect of the Congo Basin initiative and similar projects elsewhere in Africa was the degree of dependence on the external environmentalist community and the paucity of ownership or oversight by the societies on location. By contrast, KAZA was a homegrown initiative that seemed ripe for relations on an equal basis with external actors serving as collaborative partners, not just patrons.

All that said, in 2009, the future of KAZA was far from guaranteed. Political, environmental, and economic issues posed sufficiently grave problems to threaten the viability of whole enterprise. Some of the most pressing of these related to sovereignty and security, the latter of primary interest here. In the early years of the twenty-first century, southern Africa faced serious crime problems, including organized networks of narcotics traffickers and poachers.[674] Zimbabwe's political and economic devolution had flooded the region with refugees desperate for any kind of livelihood. Southern Africa also appeared to be a comfortable refuge for some international terrorists.[675] The anticipated loosening of border controls to accommodate KAZA's economic objectives boded ill for a regional response to the problem of economic refugees, criminals, and terrorists.

In early 2009, no one had yet defined the security architecture for the new arrangement, nor had the police and military establishments of the affected countries been consulted on roles they could or should play, if any. Exclusion of the security agencies was hardly surprising. African governments are not characterized by close communication between ministries and agencies. African public sectors often resemble exclusive fiefdoms, with agencies and individuals inclined to a jealous obsession with their "turf." Too, African military personnel have an unsavory reputation in much of Africa. They are widely viewed by their fellow countrymen as undemocratic, illiberal, unaccountable, uneducated, undisciplined, and inept.[676] Any large-scale presence of military officers could be expected to chill the enthusiasm of other participants. At the same time, the possibility that KAZA could be exploited by poachers, criminals, and dissidents was a serious potential threat to its long-term viability.

The very fact that the southern Africans were embarked on such a novel enterprise with so little precedent may have opened at least one significant window of opportunity. The project offered some incentive for involving the widest possible range of actors that could make a positive contribution—including, at least in theory, representatives of the security agencies. Southern African security sectors had many educated and capable officers with excellent organizing and analytical abilities, attributes often overlooked by their countrymen. KAZA seemed to offer a particularly interesting opportunity to reconnect regional militaries with their parent societies in partnerships where military personnel could play valuable niche roles.

THE REGIONAL EMBRACE OF HUMAN SECURITY

Prior to 2009, the KAZA consultations included little input from the southern African militaries, but there was a theoretical rationale in the region for doing so. Most of the southern African countries subscribed in greater or lesser degree to the broad new conceptualizations of security generally categorized under the rubric of "human security."[677] The term itself was popularized by the United Nations in the early 1990s, and the new paradigm increasingly was embedded in UN agencies and approaches.[678] The UN portrayed its approach as "people-centered" (rather than state-centered). Its most basic components were freedom from fear and freedom from want. The new UN security formula had various constituent parts, including "environmental security" that protected people from the short- and long-term ravages of nature, man-made threats in nature, and deterioration of the natural environment.

The broader definitions of security did not resonate everywhere, and some Western scholars challenged the human security paradigm as a whole, while others rejected the notion that the environment and security should be linked.[679] Not all the world's scholars were equally enthusiastic about government embrace of environmental agendas. Some were naturally suspicious of any governmental interest in the environment and worried that "securitizing environmental issues [risked] state cooption, colonization and emptying of the environmental agenda."[680] Such differences were also evident in arguments over the meaning and implications of environmental security, a construct that continues to elude a widely accepted definition.[681]

Even so, African scholars were among the early supporters of the new "security" thinking and tended to endorse the broader definition exemplified in the UN conceptualization. By the 1990s, human security themes had become increasingly prominent in the thinking of African officials and academics.[682] Interestingly, "environment" appeared in most of the new African definitions of security, either in terms of a human right to a healthful environment or in terms of rights of common citizens to environmental resources.[683]

Initially, the broader models of security did not leave much room for the coercive agencies of the state and seemed to deny security officials any exclusive right to define the subject. The new thinking brought into question the relevance of the traditional military establishments themselves. However, a number of countries soon adapted the broader security models to military roles and missions. One of the most remarkable of these was postapartheid South Africa. After 1994, that country endeavored to realign its military to its new national priorities and commitment to human security.[684] South Africa's foundational document for its new military establishment articulated a whole new philosophy of national defense, capturing one of the most expansive definitions of security on record anywhere and offering a clear environmental dimension:

> National security is no longer viewed as a predominantly military and police problem. It has broadened to incorporate political, economic, social and environmental matters...security is an all-encompassing condition in which individual citizens...inhabit an environment which is not detrimental to their health and well-being.[685]

The linkages in the South African White Paper were not foreign to officials in other African countries. South Africa's new approach complemented, for instance, Botswana's use of military power to pursue environmental ends in antipoaching.[686] By the twenty-first century, southern Africans had established an unambiguous conceptual relationship between militaries and environmental security in their region, though they were far from a full exploration of its possibilities.

THE REGIONAL MILITARY ANGLE

Whether or not the KAZA project ever succeeded as envisioned by its proponents, Botswana could take credit for much of its inspiration. That country's prolific wildlife sanctuaries were at the very center of the project. In one sense, the whole KAZA enterprise could be viewed as an attempt to extend the commercial success of Botswana's wildlife-based tourism industry to the entire southern African region. And, Botswana also was one country that had pursued environmental security through resorting to military force, having deployed its defense force into its wildlife conservancies since 1987 in a successful, long-term effort to halt egregious, commercial megafauna poaching.

The story of military antipoaching in Botswana is the account of a government seeking to safeguard its natural resources and its population from vicious assault. In the mid-1980s, Africa's elephants and rhinos were severely threatened. Networks of well-armed criminals with links to the Persian Gulf and Far East sponsored much of the slaughter. The criminal gangs were patient and persistent and could easily outwait (and usually outwit) the sporadic responses of their African public sector opponents. They were active in virtually every African location that harbored the large animals.[687]

The commercial poachers not only threatened Botswana's wildlife and tourist industry, they assaulted and robbed local citizens living near the wildlife conservancies. The country deployed its military into the national parks to confront the poachers in 1987, but that decision was fraught with the real possibility of failure. The small defense force, a mere decade old at the time, had not been created or trained for an anti-poaching role. The handful of countries in Africa that previously had confronted poaching with military force had seen virtually no success. Even so, despite the difficulties and uncertainty at the outset, Botswana's antipoaching operations succeeded brilliantly, largely halting the poaching and incurring the approval of the general population.[688]

Botswana's accomplishment has not been universally applauded. Some environmentalists interviewed by the author strongly commended Botswana's initiative and praised the performance of the defense force, while others were much less sanguine. None condemned the usage outright, and none denied that it had prevented at least some poaching. But scholars familiar with Botswana's antipoaching activities were ambivalent about the wisdom of "protectionist" conservation programs wherever found, including Botswana. The same was evident in the KAZA consultations though 2008. Local environmentalists, members of nongovernmental organizations, and civil servants were leery of military involvement and skeptical of its merits, however well intended.

Successful use of military force in an environmental security role by one African country does not validate that usage as a universal norm, of course. Nor does it even prove that military deployment is the best solution to the problem of commercial megafauna poaching. However, it does suggest that African security agencies can play useful environmental security niche roles in carefully defined circumstances and should not be excluded from the discourses on the topic. What was missing in KAZA in 2009 was a dialogue between environmentalists, civil government officials, and military leaders to explore the limits of the appropriate and the art of the possible in addressing the project's security issues and other niche roles that militaries might play.

Beyond Botswana and its antipoaching operations, the only southern African country with a serious military involvement in environmentalism in 2009 was South Africa. That country maintained a small environmental office within its military headquarters, concerned with a broad range of environmental issues, although its focus was limited almost exclusively to military installations and their immediate environs. The South Africans also employed their air force in routine coastal patrolling to secure their maritime resources, drawing aircraft from a squadron based in Cape Town and linking that to coastal law enforcement. Just as significantly, they served as a bridge to military environmental activities in other African countries, hosting consultations on environmental issues among military officials. These initiatives demonstrated that African militaries themselves were interested in exploring possible roles in regional environmental issues.[689]

EXTERNAL PARTNERSHIPS IN GENERAL

A rich assortment of environmental linkages extends from southern Africa to a global community. Local scholars have long participated in the worldwide debates linking environment and security.[690] Southern African academe and civil society groups have served as a vital bridge to the larger world. The world community, in turn, has exerted a significant impact on conceptions and processes of conservation in southern Africa. The results include bilateral and multilateral public sector partnerships. Results also are evident in international agreements such as the Convention on International Trade in Endangered Species (CITES) or the Kyoto Protocol on greenhouse gases. Other linkages bind southern African countries to international environmental experts and advocacy communities. Both the local governments and advocacy groups in southern

Africa work with organizations like the United Nations Development Programme (UNDP), WWF, IUCN, and Conservation International.

In 2009 southern Africa's government officials, local scholars, and civil society activists were entirely conversant with the debates current elsewhere in the world. However, very few of these connections had any military dimension, and most of the few military-environmental connections that did exist involved the United States. U.S. military association with southern Africa's environmental issues had been episodic and generally bilateral. They did not constitute a particularly good transregional model.[691] However, this was about the only external military environmental game in town, so it warrants attention.

In fact, U.S. military partnerships on environmental issues were at something of a low point in southern Africa in 2009, having seen a higher intensity in the 1990s. That earlier activity largely involved the military "biodiversity" programs intended to encourage African armies to broaden their focus into environmental protection.[692] In 2009 the only more-or-less permanent military environmental connection in the region was U.S. participation in an environmental security working group with South Africa.[693]

There were, however, indications that U.S. military interest was on the upswing. One was the activity of the Africa Center for Strategic Studies, a U.S. Department of Defense institution formed in 1999 with the role of bringing senior civil and military officials from across the continent together in a neutral, academic setting to analyze issues of civil-military relations, security sector resourcing, and formulation of national security strategy. This center subsequently proved to be an excellent vehicle for communication among midlevel and senior African civil government and security sector officials. By 2006, it had sponsored colloquia and research with environmental themes and seemed poised to increase that focus in the future.[694]

A second indication was the U.S. government announcement, in early 2007, that it was creating a new military Africa command (AFRICOM) with exclusive responsibility for overseeing U.S. security interests in Africa. By late 2008, this command had achieved initial operating capacity and was making vigorous efforts to connect with partners in Africa. U.S. officials were going to considerable lengths to highlight the uniqueness of this new command. It would reflect a strong "interagency" complexion, heavily staffed by personnel from nonmilitary executive branch agencies. It would focus much more on "soft power" (humanitarian issues) than equivalent U.S. military commands elsewhere. Insiders suggested it would concern itself with regional issues like human development and environmental security.[695]

In 2009 Africans, for their part, still exhibited considerable ambivalence about the advent of the AFRICOM. Some expressed fear of American hegemonic intentions and the prospects that Africans would be unwillingly entangled in the U.S. global War on Terror.[696] Others questioned the propriety of using a military organization to pursue "soft-power" objectives.[697] These African concerns seem to suggest that if AFRICOM succeeds in forging a real partnership with African countries, that outcome will come slowly and will derive from cooperation over issues that tend to unite rather than divide. Environmental security may offer exactly that kind of bridging issue.

The U.S. military itself has a rather checkered history when it comes to the environment, but its interest in environmental issues has increased since the early 1990s, evident in organizations such as the Army Environmental Policy Institute (AEPI) and the Army Strategy for the Environment.[698] There is considerable room for the environmental community to encourage that interest, which could have a potentially positive impact on U.S. regional engagements. Organizations like the Woodrow Wilson International Center's Environmental Change and Security Program (ECSP) that reach out to military audiences are good examples of what can be done.[699]

The U.S. military is likely to remain the external military actor with the most widespread African involvements and the most resources to bring to any regional security issue. It has engaged in limited environmental activity in Africa over the past two decades and has yet to demonstrate that it has the incentive or vision to make a significant difference on the environment, but that may change. KAZA offers a bellwether opportunity.

DIMENSIONS OF A NEW ERA OF OPPORTUNITY

Africa is one region of the world where threats emanating from the natural environment are dire and lethal and where the capacities of local human communities to deal with the effects often are very limited. This also is a region where resource scarcity (particularly water) may strain future intercommunity relations and lead to serious conflict. At the same time, it is a region rich in biodiversity, yet its natural environment, a precious heritage of all mankind, is facing serious threat. Southern Africa is home to dedicated environmentalists seeking to be good stewards of their continent's environmental riches. Their governments, now more than ever, are interested in partnerships with external parties to address these environmental problems.[700] The emergence of the African Union in the early years of the twenty-first century was a hopeful sign that African leaders would be more willing to deal with difficult interstate and interregional issues such as the environment.[701]

The same points could be made about subregional organizations such as the Southern African Development Community (SADC). These have become much more focused on security issues since the early 1990s and have increasingly displayed an interest in broad problems of conflict or humanitarian crisis, including issues of environmental security.[702] African countries are interested in finding mechanisms for attenuating environmental crises and other natural disasters. Regional organizations offer opportunities for collaboration on issues of environmental security, often when states have difficulty agreeing on other issues. The same is true of projects like KAZA. Now, particularly, is a time when perceptive external partners may be wise to view environmental security in Africa as a "peace multiplier."[703]

The author's interviews with a wide range of interested individuals in southern Africa between 2005 and 2008 suggested that military establishments could play a useful role in emerging environmental partnerships. While not all were convinced that military planners should participate in the KAZA consultations, interviewees called attention to the unique capacities of the military in organizing responses to environmental catastrophe. Several were familiar with the U.S. Army Corps of Engineers and suggested that its expertise could be useful to river-system management. Interviewees frequently expressed curiosity about the contribution that military experts could make. Several organizations seemed willing to host such consultations.[704]

The appropriate role for the military in environmental partnerships remains a very open question, but southern Africa appears to be a promising locale for their examination and experimentation. Government capacity is greater in this region than in much of the rest of the continent, and southern African governments are committed by public doctrine to environmental stewardship. Compared to the rest of Africa, civil society in this region also contains an unusually committed and capable environmentalist constituency. More to the point, southern African countries have just begun to engage in extraordinary multilateral ventures that link human development, economic growth, and the environment and in which the security ramifications have yet to be defined. If these initiatives succeed, they could provide a powerful model for the war-ravaged economies further north. It is a propitious time for external partners to support these efforts in a variety of collaborative partnerships, including, potentially, military ones.

IMPLICATIONS FOR EXTERNAL PARTNERS

"Environmental" issues in Africa consistently overlap other important objectives such as health, infrastructural rehabilitation, and economic development. A holistic approach, embedded in genuine, wide-ranging regional partnerships, offers a significant new option to promote and sustain regional stability. KAZA could be at the center of this approach, at least in the southern Africa region. External partners could play a constructive role.

For security issues, this could take the form of encouraging dialogue to define the appropriate military contribution. That role could profitably emphasize the mediation of local institutions (such as the South African Institute for Security Studies or the Graduate School of Public and Development Management at the University of Witwatersrand) well placed to serve as facilitators of civil-military dialogue.[705] An important part of the dialogue would be the linkage of currently unconnected stakeholders (civil servants, military officers, environmentalists, scientists, entrepreneurs, and prospective foreign partners) to a common vision for environmental security. Such a process probably also would create bridging opportunities for collaboration on other issues, such as peacekeeping and humanitarian relief.

The importance of genuine partnership cannot be overemphasized, and requires wisdom on the part of officials in countries outside the African continent. Because of the recent history of the region, Africans easily find offense in patronizing attitudes by outsiders who offer "solutions" to African problems without really understanding African conditions, a circumstance doubly annoying to Africans when prospective partners ignore African institutions already engaged in the issues. Yet, outside assistance can play a useful role in helping Africans with homegrown solutions to their own economic, environmental, and security dilemmas. KAZA represents a good opportunity to explore the bounds of such partnerships.

CHAPTER 11

CONCLUDING REMARKS

ABOUT

ENVIRONMENTAL AND HUMAN SECURITY

AND THE ROLE OF U.S. MILITARY FORCES

IN AFRICA IN THE FUTURE

Maxie McFarland and Robert Feldman[706]

INTRODUCTION

Albert Einstein once said, "Everything should be as simple as possible, and no simpler." In essence, Einstein postulated that there comes a point in defining problems when further reduction to lower common denominators becomes counterproductive. This thought lends itself to the theory that the key to solving complex problems is in finding the critical point just below a degree of complexity that is unmanageable and just above a degree of simplicity that loses context with the problem.

Unfortunately, in the case of Africa, an approach as simple as possible and no simpler still leaves you with a very complex, contradictory, compounding, and competing set of factors that defy reduction to manageable states. This leads to a state where those wanting to help either end up confused and overwhelmed by the complexity or disillusioned by a default to simplicity that serves no long-term purpose.

Africa security issues contain a complex set of interconnected problems and opportunities so intertwined that isolation is not possible. Conversely, it is absolutely necessary to subdivide these complex problems into their various elements and to determine the linkages between them as a way of understanding the problem and assessing opportunities as a whole. Analysis, therefore, is only a means of reaching synthesis with respect to defining the problem and thinking holistically about its solution. The authors in this book have done a marvelous job of analyzing both problems and opportunities in Africa and synthesizing them into holistic perspectives of activities that have sound prospects for a long-term positive impact. While the military's role in these activities certainly should be explored, in the end, the military is not the sole solution to Africa's problems, and AFRICOM should not be viewed in that light.

Without a doubt, AFRICOM has been created as a different kind of military command with an appropriately different focus. Unlike other combatant commands, AFRICOM has a focus broader than military operations and incorporates diplomatic, economic, and humanitarian aid to support conflict prevention. It was established with a clear understanding that military operations and capabilities are not the sole solution to Africa's problems. Instead, AFRICOM seeks to bring military strengths in planning, coordination, execution, and integration to bear in support of operations by NGOs, African states, regional organizations, and the U.S. Department of State to help provide developmental assistance. AFRICOM represents a shift to a more long-range and enduring military commitment, as it seeks to enable both state and human security efforts across the continent. Though civil affairs, engineers, medical, and other

units have performed missions in Africa before—digging wells, vacci-
nating children, and providing other important services—the emphasis
on the military to partner in numerous long-term development projects
in Africa is fairly new.

All human activity is conducted within the context of its environment.
It is impossible to separate economic, health, safety, social, governance,
or any issue associated with either state or human security from that
of the environment. Consequently, environmental security will play a
prominent role in any strategy to address problems in Africa. Programs
dealing with conflict, famine, poverty, disease, and other serious problems must also address the environmental factors such as natural resource
use, conservation, and sharing in order to be effective. The papers presented at the ISA convention examine these relationships, providing
detailed examples of not just how environmental projects impact human
security, but also the myriad problems that frequently occur when under-
taking them in Africa. Reviewing the findings from these examples, the
papers' authors offer recommendations as to how AFRICOM might help
in planning, implementing, and supporting the management of environmental projects, including ways to increase the chances during the planning stage of discovering potential unwanted second- and third-order
effects so they can be avoided.

AFRICOM's biggest issues lie in the areas of suspicion and expectation. NGOs, environmentalists, many African states, and international and regional organizations view the military with a great
deal of suspicion and skepticism. African history does not provide a
foundation of trust between local communities and military organizations. NGOs question the military motives, their long-term commitment, and their neutrality in dealing with human security issues. On
the other hand, these same organizations often expect the military
to achieve immediate results with their "vast resources" and disci-
plined approach to getting things done. Most NGOs do not jump willingly into partnerships with the military except in cases of extreme
humanitarian crisis. With respect to environmental security, these

organizations express many of the following concerns related to a possible role for the military.

- Securitizing environmental issues risks having these co-opted by the state.
- The relevance of traditional military establishments in dealing with environmental issues is questionable, since issues might best be served by reinvesting dollars spent on military capability into civilian projects focused on the environment.
- Preparedness for the performance of tasks associated with environmental projects is doubtful because soldiers are not taught how to deal with environmental problems.
- Armies in some countries have a well-deserved reputation for human rights violations and can be extremely difficult to convince indigenous people to work with them.
- Some NGOs have negative prior experiences when dealing with militaries on issues not related to the environment and previous interactions color the perspective of the organizations regarding possible future working relationships.

The reluctance to at least engage military planners in the development of major environmental or economic projects often leads to significant downstream issues in the complex African environment. As an example, Dan Henk describes in his chapter the Kavango Zambezi (KAZA) Transfrontier Conservation Project. This failure to involve security agencies during the planning stage essentially ensured there would be problems later on, as the project progressed.[707] Postponing questions of how to deal with poachers put the entire project's viability at risk.

Fortunately, the tide appears to be changing. Despite the misgivings of some, at least a few African militaries have begun to venture into environmental security. Botswana is a prime example. During the mid-1980s, well-armed criminals, many with links to the Persian Gulf and Far East, engaged in large-scale slaughter of elephants and rhinos, threatening the country's wildlife and tourist industry. The outgunned and outmaneuvered park rangers were largely ineffective. However, the Botswana Defense Force—although small, only ten years old, and not trained to deal with poaching—intervened and was extremely successful at curbing the slaughter.[708]

AFRICOM's Role in Human and Environmental Security

There are potentially five broad roles AFRICOM could play in response to environmental and human security problems in Africa.

- *Planning:* The U.S. military's ability to plan and conduct complex, long-term campaigns is among the best in the world. U.S. military planners are sought after by non-DoD agencies and are key components to many non-military efforts in the U.S. government. AFRICOM could play a leading role in developing long-term plans in support of operations that are not inherently military in nature. Its ability to consider second- and third-order consequences; piece together sequential, complementary, and supporting activities; and integrate all available capabilities oriented on desired outcomes could provide the foundation for many long-term regional projects throughout Africa.
- *Coordination:* As a permanent headquarters, AFRICOM, with its USAID and regional partnerships, offers great potential as the single coordination and deconfliction node. New NGOs, business enterprises, and humanitarian organizations arrive on the scene constantly. As a central repository for ongoing operations, AFRICOM could provide key information on other operations that might be complementary to those being considered by these new organizations. It could also coordinate U.S. military operations to deconflict activities to achieve the most effective use of often-limited resources. AFRICOM can facilitate a neutral environment for civilian and military leaders from different countries to come together to discuss transnational issues such as the formation of transfrontier parks.

The Africa Center for Strategic Studies is an example of success in this area. Gentle encouragement and guidance from AFRICOM can result in long-term benefits for all involved.

- **Enabling:** AFRICOM can play a key role in enabling local security in support of human security and environmental activities by nonmilitary organizations. This entails training and equipping local security forces, providing reinforcement when they are over-whelmed, and, when properly coordinated and requested, providing the actual local security necessary for these microenterprises to flourish. An excellent example of this already exists. Combined Joint Task Force-Horn of Africa (CJTF-HOA) has been assisting pastoralists in Kenya and Ethiopia. In addition to providing security, it has developed well-received water, school, and health clinic projects along strategic transit routes from those two countries into Somalia. Even Somalis inside Somalia have heard of the good works the Americans are doing.[709]

- **Deterring:** Conflict often leads to environmental degradation through mechanisms such as the creation of large numbers of refugees who crowd into camps on marginal lands. Deterring or ending conflict could potentially bring about environmental gains. It is uncertain whether peace and subsequent human security can be obtained unless significant military might is brought to bear in the Democratic Republic of the Congo, Darfur, Somalia, or other places on the African continent where conflicts rage. Indeed, this paper does not advocate for or against American military intervention in any of those areas. However, it is necessary to recognize that the United States could play a role with direct support using American troops and with indirect support by providing advisors, equipment, and planning capacity for African militaries. If directed by policy makers, a combination of all of these capabilities could be employed, and substantial environmental benefits could accrue from the cessation of these conflicts.

- **Partnering:** AFRICOM can serve as a partner where the military's expertise and capabilities complement those of aid organizations.

Perhaps a Peace Corps volunteer identifies a village that would greatly benefit from a well, but the water table is fairly deep. Since the Peace Corps lacks both the drilling expertise and the equipment, AFRICOM could provide these for a successful well-drilling operation.

These five roles are not mutually exclusive, nor are they always clearly demarcated. The soldier patrolling a hostile area, during breaks from fighting, might help construct a local school; the soldier assigned to provide security for an NGO might be temporarily relocated to another part of the country to assume a more active combat role. Soldiers will need to be adaptable to sudden or opportunistic changes in missions.

Conclusion

The U.S. military is not *the* solution to Africa's problems. As stated previously, Africa presents a significant set of interconnected problems, resulting in a framework of organized complexity requiring holistic and synergistic solutions. However, the U.S. military, and in particular, AFRICOM, can play a pivotal role in environmental security projects. The question becomes the nature and scope of that pivotal role. There is a big difference between what the military can do, what it should do, or even what it will be allowed to do. The opportunities offered by an organization like AFRICOM can be easily marginalized if it is not advantaged or if it is advantaged inappropriately. As George Bernard Shaw once wrote, "To be in hell is to rift, to be in heaven is to steer." How should we steer AFRICOM?

Although the environment and security are significant issues, it remains to be seen whether they represent the correct lowest common denominator of simplicity and complexity for defining and solving African problems. Historically, few problems in Africa were effectively isolated and solved over the long term. In fact, solutions to isolated problems, in most cases, resulted in the creation of new problems or an exponential expansion of existing problems. Thus, there is a need for holistic and comprehensive approaches that exceed what the military can or should bring to the table.

Stability and prosperity in Africa is an important global issue and one that is key to the United States as well. When people are desperate for assistance, radical organizations such as Al-Qaeda will step in and fill the gap. When an African government participates in successful environmental projects, such as raising more food crops or providing potable water, it fulfills a key need for the population. This success also increases the government's legitimacy among the citizens, helping to turn them away from radical organizations that might threaten the United States.

The papers presented at the ISA convention describe several environmental projects, warts and all. Clearly, a large number of lessons were learned the hard way. Military planners need to learn from these studies as they look at the development of long-term campaigns to address U.S. interests on the African continent.

It is interesting to note that in all of those papers, the authors tend to agree that AFRICOM can be useful but should not lead the way, and rather, should help Africans develop their own homegrown solutions to their environmental and security problems. Only in this manner can Africans find the appropriate long-term strategies to overcome some of the enormous difficulties they face.

As AFRICOM, NGOs, states, businesses, and international and regional organizations navigate the difficult waters of cooperation and partnerships, there are several factors that all should keep in the back of their minds:

- Although there are opportunities for AFRICOM to improve environmental security in Africa, there are also chances that second- and third-order effects from well-intentioned actions could be counterproductive. Any planning must attempt to understand all possible consequences from AFRICOM interventions.
- New or redesigned training programs will be required for American soldiers to learn how to effectively work in the areas of environmental security in Africa. Courses in languages (including some

rather remote tribal dialects), culture, and using sustainable and appropriate technologies will need to be offered alongside combat courses that emphasize the challenges of fighting in Africa.

- AFRICOM should learn from other organizations who have been in Africa for a long period, including the UN, NATO, and numerous NGOs.
- Global climate change, which will potentially impact Africa more than the other populated continents, will accelerate both Africa's problems and the need to find solutions to them. Implementing projects to mitigate the negative impacts of global climate change should be an AFRICOM priority.
- There are numerous obstacles to success in Africa, with poor governance perhaps being the most difficult to overcome. AFRICOM will have to develop successful strategies to work within the various frameworks of existing African governments.
- With only limited resources, AFRICOM will have to choose projects carefully. Using appropriate metrics will be critical to deciding which to pursue.
- Because so many environmental problems in Africa have numerous underlying causes, AFRICOM will sometimes need to look for feasible solutions as opposed to optimal.
- The way ahead for many African problems is through education of women. However, this is often a veritable minefield of controversy.
- What happens in Africa does not stay in Africa. Refugees, terrorists, windblown dust, and diseases will find their way to other continents. AFRICOM's work is not just helping Africa, it's helping everyone.
- Many parts of Africa are wary or outright hostile to AFRICOM. This military command will have to strive to improve its relations with the peoples of Africa.
- Success is not guaranteed in any of the endeavors. Even when it appears in hand, an unexpected coup or other situation-changing event could suddenly occur. AFRICOM will need to be resilient: accept when failures occur, learn from its mistakes, and carry on.

What will AFRICOM most need to transform in order to succeed in its role of helping Africans attain environmental and human security? Jebb and her coauthors state,

> For the U.S. military that is currently transforming, it is clear that the most important asset that requires transforming is the intellect. The military cannot transform effectively without ensuring proper education and training of its service members and true "buy in" to the human components of human security by its leaders. The foreseeable future will require adaptable, innovative, and culturally aware military members.[710]

In conclusion, there is definitely a role for AFRICOM to play in environmental security in Africa. The debate over what that role should be should not stop the command from taking action on issues as identified. In many cases, it is only through taking action that a problem becomes clarified. The challenge will be to choose the right set of actions that will serve to clarify problems, solutions, and roles.

NOTES

PREFACE

1. Dr. Turton notes in his analysis that the collapse of water sanitation controls since 1994 is linked to the government's lack of capacity and shortage of experienced water purification professionals in the public sector. There is currently a widespread shortage of experienced scientists and engineers in many fields in South Africa. Since 1994, many experienced non-African scientists and engineers have left South Africa to find jobs abroad. The way affirmative action programs are implemented is often cited as having contributed to this exodus and the resulting skill shortage. Critics of current South African affirmative action programs label these practices reverse racism. For an example of this perspective, see R. W. Johnson, *South Africa's Brave New World: The Beloved Country Since the End of Apartheid* (London: Allen Lane, 2009).

2. Anthony Turton, *Three Strategic Water Quality Challenges that Decision-Makers Need to Know About and How the CSIR Should Respond* (A Clean South Africa), planned keynote address at the CSIR conference, "Science Real and Relevant," Pretoria, South Africa, November 18, 2008. Council for Scientific and Industrial Research. Available at http://www.environment.co.za/documents/water/keynoteAddressCSIR2008.pdf.

CHAPTER 1

3. The chapter is a slightly revised version of "Human and Environmental Security in the Sahel: A Modest Strategy for Success," in *Environmental Change and Human Security: Recognizing and Acting on Hazard Impacts*, ed. P. H. Liotta, D. A. Mouat, W. G. Kepner, and Judith M. Lancaster, 341–392 (Dordrecht, The Netherlands: Springer, 2008). The work is part of the NATO Science for Peace and Security Series C: Environmental Security, reprinted with the permission of the editor and Springer. The original publication is available at http://www.springer.com.

4. The views expressed are those of the authors and not necessarily those of the U.S. Military Academy or the Department of Defense.

5. The TSCTP is an attempt to assist states of the Sahel combat terrorism by providing both military and nonmilitary assistance.

6. Peter Senge, *The Fifth Discipline: The Art & Practice of The Learning Organization* (New York: Currency Doubleday Books, 1990), 73. Senge coined the term "a shift of mind." Shifting our Newtonian understanding of cause and effect to a systemic view allows people to understand that the whole is greater than the sum of its parts and to understand the interrelationships of entities.

7. Fritjof Capra, *The Web of Life: A New Scientific Understanding of Living Systems* (New York: Anchor Books Doubleday, 1996), 25.

8. Ibid., 31. According to Capra, Christian von Ehrenfels, a philosopher, coined the phrase "the whole is more than the sum of its parts." This became the basis of all systems-thinking approaches.

9. Robert Jervis, *System Effects: Complexity in Political and Social Life* (Princeton, NJ: Princeton University Press, 1997), 12–13. Further, Jervis believes, "If we are dealing with a system, the whole is different from, not greater than, the sum of the parts." This counters Senge and Capra's view that "the whole is greater than the sum of its parts." The differences in opinion still recognize that the parts of a system only present half of the integrated whole.

10. Capra, *The Web of Life*, 158, 160, and 209. In Capra's book *The Hidden Connections: Integrating the Biological, Cognitive, and Social Dimensions of Life into a Science of Sustainability* (New York: Doubleday, 2002), he refines his universal living systems theory as he integrates the social and human domains of a living system inherent to a human system. He adds the criterion of meaning.

11. James G. Miller, *Living Systems* (New York: McGraw-Hill Book Company, 1978), 1. Also refer to table 1-1, the nineteen critical subsystems of a living system, on p. 3.

12. Capra, *The Hidden Connections*, 81.

13. See Cindy Jebb and Madel Abb, "Human Security and Good Governance: A Living Systems Approach to Understanding and Combating Terrorism" (INSS, Colorado Springs, CO, December 31, 2005).

14. Capra, *The Hidden Connections*, 73.

15. Ibid., 83–87.

16. Francis Fukuyama, *State-Building: Governance and World Order in the 21st Century* (Ithaca, NY: Cornell University Press, 2004), 26.

17. Note that this section on human security is modified from Lianne Kennedy and Cindy R. Jebb, "Non-state Actors and Transnational Issues," in

American National Security, ed. A. Amos, A. Jordan, William J. Taylor Jr., Michael J. Meese, and S. C. Nielsen (Baltimore: Johns Hopkins University Press, forthcoming).

18. United Nations General Assembly, "Part One: Towards a New Security Consensus," *Report of the High-Level Panel on Threats, Challenges, and Change to the Secretary General*, http://www.un.org/secureworld/report.pdf.

19. See Mohammed Ayoob, *The Third World Security Predicament* (Boulder, CO: Lynne Rienner, 1995); Barry Buzan, *People, States, and Fear*, 2nd ed. (Chapel Hill: University of North Carolina, 1983); Charles Tilly, "War Making and State Making as Organized Crime," in *Bringing the State Back In*, ed. Peter B. Evans, Dietrich Rueschmeyer, and Theda Skocpol (Cambridge: Cambridge University Press, 1985), 169–187; Mostafa Rejai and Cynthia H. Enloe, "Nation-States and State-Nations," *International Studies Quarterly* 13, no. 2 (June 1969): 140–158.

20. *United Nations Development Programme Report* (1994), 3, 22–23, as quoted by Peter H. Liotta, *The Uncertain Certainty* (Lanham, MD: Lexington Books, 2004), 4–5.

21. *A Human Security Doctrine for Europe, The Barcelona Report of the Study Group on Europe's Security Capabilities*, September 15, 2004, 9.

22. United Nations General Assembly, "Part One: Towards a New Security Consensus."

23. Ibid., 22. Also see Cindy R. Jebb, P. H. Liotta, Thomas Sherlock, and Ruth M. Beitler, *The Fight For Legitimacy: Democracy Versus Terrorism* (Westport, CT: Praeger Security International, 2006), 134–136.

24. See, for example, Morton H. Halperin, "Guaranteeing Democracy," *Foreign Policy* 91 (Summer 1993): 105–122 and Lee Feinstein and Anne-Marie Slaughter, "A Duty to Prevent," *Foreign Affairs* 83, no. 1 (January/February 2004): 136.

25. Also see Tedd Gurr, "Why Minorities Rebel: Explaining Ethnopolitical Protest and Rebellion," in *Minorities and Risk: A Global View of Ethnopolitical Conflicts*, (Washington, DC: United States Institute of Peace: October 1997), 123–138 and Robert Kaplan, "The Coming Anarchy," *The Atlantic Monthly* 273, no. 2 (February 1994): 44–76.

26. As cited in Peter H. Liotta and Taylor Owen, "Sense and Symbolism: Europe Takes on Human Security," *Parameters* 36, no. 3 (2006): 90.

27. These goals are taken from UNDP *Human Development Report, 2005: International Cooperation at a Crossroads: Aid, Trade and Security in an Unequal World*, 39.

28. Jan Pronk, "Globalization, Poverty, and Security," in *Human and Environmental Security: An Agenda for Change*, ed. Felix Dodds and Tim Pippard, 84 (London: Earthscan, 2005).
29. UNDP, *Human Development Report, 2005*, 37–38. The only country that measures a greater gap than the global scale between the very rich and the very poor is Namibia.
30. Jeffrey D. Sachs, *The End of Poverty: Economic Possibilities of Our Time* (New York: Penguin Books, 2005), 192.
31. Ibid., 194.
32. Ibid., 195.
33. UNDP, *Human Development Report* (1994), 35–37.
34. H. B. Cavalcanti, "Food Security," in *Human and Environmental Security: An Agenda for Change*, ed. Felix Dodds and Tim Pippard, 156 (London: Earthscan, 2005) and Lester R. Brown, *Outgrowing the Earth: The Food Security Challenge in an Age of Falling Water Tables and Rising Temperatures* (New York: W. W. Norton and Company, 2004), 178 and Lester Brown, *Outgrowing the Earth*, 178.
35. Cavalcanti, "Food Security," 163.
36. J. H. Bodley, "Evolution of Food Systems, " in *Anthropology and Contemporary Human Problems*, ed. J. H. Bodley (Mountain View, CA: Mayfield), 83–84.
37. Ibid, 83–122.
38. Cavalcanti, "Food Security," 153–154.
39. Lester R. Brown, *Outgrowing the Earth: The Food Security Challenge in an Age of Falling Water Tables and Rising Temperatures* (New York: W. W. Norton and Company, 2004), 4–9. See Cavalcanti reference the effects of shocks in "Food Security," 154.
40. Brown, *Outgrowing the Earth*, 19.
41. Stefan Elbe, "HIV/AIDS and the Changing Landscape of War," in *New Global Dangers: Changing Dimensions of International Security*, ed. Michael E. Brown, Owen R. Cote Jr., Sean M. Lynn-Hones, and Steven E. Miller, 371 (Cambridge, MA: MIT Press, 2004).
42. Nicholas Eberstadt, "The Future of AIDS," *Foreign Affairs* 81, no. 6 (November/December 2002): 42.
43. Christine K. Durbak and Claudia M. Strauss, "Securing a Healthier World," in *Human and Environmental Security: An Agenda for Change*, ed. Felix Dodds and Tim Pippard, 128–129 (London: Earthscan, 2005).
44. Ibid., 129.

45. Ibid., 129–130.
46. Bruce Hoffman, "Terrorism" (lecture to Terrorism and Counterterrorism class, U.S. Military Academy, West Point, NY, April 2004).
47. *Arab Human Development Report 2002: Creating Opportunities for Future Generations* (New York: United Nations Development Programme), 27, as quoted in Ruth M. Beitler and Cindy R. Jebb, "Egypt as a Failing State: Implications for U.S. National Security," *INSS Occasional Paper* 51 (July 2003): 47.
48. Thomas M. Sanderson, "Transnational Terror and Organized Crime: Blurring the Lines," *SAIS Review* 24, no. 1 (2004): 2.
49. Ibid.
50. R. Dixon, "The World; Secret Lives of Servitude in Niger; The Government Has Banned Slavery and Denies It Exists. Though Few Speak of It, the Practice Is a Tradition Many Do Not Question," *Los Angeles Times*, September 3, 2005, A2–A3.
51. Reven Paz and Moshe Terdman, "Africa: The Gold Mine of Al Qaeda and Global Jihad," *Global Research in International Affairs Center, the Project for the Research of Islamist Movements, Occasional Papers* 4, no. 2 (June 2006): 2.
52. Lester R. Brown, *Plan B 2.0: Rescuing a Planet under Stress and a Civilization in Trouble* (New York: W. W. Norton, 2006), 3.
53. Thomas F. Homer-Dixon, "Environmental Scarcities and Violent Conflict," in *New Global Dangers: Changing Dimensions of International Security*, ed. Michael E. Brown et al., 265 (Cambridge, MA: MIT Press, 2004).
54. Ibid., 267.
55. Michael T. Klare, *Resource Wars: The New Landscape of Global Conflict* (New York: Henry Holt and Company, 2002), 8–9. See Klare for further discussion on resource-driven conflicts.
56. Homer-Dixon, "Environmental Scarcities and Violent Conflict," 278.
57. *Background Note: Niger*, U.S. Department of State, Bureau of African Affairs, April 2006, http://www.state.gov. See also Mark Wentling, "Draft Paper; Elements of an Assistance Strategy," June 24, 2006. Some scholars would argue that the doubling time is even closer to twenty-two years.
58. *Background Note: Niger*, U.S. Department of State, Bureau of African Affairs, April 2006, http://www.state.gov.
59. *Background Note: Niger*, 3–4.
60. Interview with Nigerien official, Niamey, Niger, August 27, 2006.

61. Exclusive Analysis, *Global Risk Outlook* (London: Author, 2005).
62. *Global Risk Outlook 2006*, 61.
63. Interview with Vanessa Ndiaye, Danish Corporation, Niamey, Niger, August 27, 2006.
64. Ousseini Issa, "Niger: Forest Squatters Demand New Homes Before Eviction," Global Information Network, August 3, 2006, 1. Available at http://0-proquest.umi.com.usmalibrary.usma.edu.
65. Interview with Mark Wentley, USAID, Niamey, Niger, August 28, 2006.
66. Mamoudou Gazibo, "Foreign Aid and Democratization: Benin and Niger," *African Studies Review* 48, no. 3 (2005): 8–9.
67. William F. S. Miles, "The Niger We Should Know," *Boston Globe*, August 23, 2005, A15.
68. Interview with World Health official, Niamey, Niger, August 28, 2006.
69. Miles further explains how women are absent in Islamic decision-making processes, thus putting into question the volunteerism of supporting radical Islam. All people we interviewed discussed the importance of education for girls as the key to development, as will be explained further in the chapter.
70. William F. S. Miles, "Shari'a as De-Africanization: Evidence from Hausaland," *Africa Today* 50, no. 1 (2003): 5.
71. Interview with U.S. Embassy official, Niamey, Niger, August 29, 2006.
72. The quote is taken from Office of Dutch-Nigerien Cooperation, "Study of the Practices of Islam in Niger," *Provisional Report*, April 2006.
73. Miles, "Shari'a as De-Africanization," 5.
74. Ibid. For a comprehensive analysis on the different forms of Islam and Islamic associations by region, see Office of Dutch-Nigerien Cooperation, "Study of the Practices of Islam in Niger."
75. Ricardo Laremont and Hirach Gregorian, "Political Islam in West Africa and the Sahel," *Military Review* 86, no. 1 (January/February 2006).
76. The existence of north Africans fighting in Iraq on behalf of Al-Qaeda is taken from Craig Whitlock, "Terror Group Expands in N. Africa; Faction Backed by al-Qaida Runs Training Camps in the Region," *Houston Chronicle*, October 6, 2006, 19.
77. International Crisis Group, "Islamist Terrorism in the Sahel: Fact or Fiction," *Africa Report* 92, no. 31 (March 2005): 7–8, 22.
78. Stephen Ellis, "Briefing: The Pan-Sahel Initiative," *African Affairs* 103, no. 412 (2004): 463.
79. Brennan Kraxberger, "The United States and Africa: Shifting Geopolitics in an 'Age of Terror,'" *Africa Today* 52, no. 1 (2005).

80. Interview with WFP personnel, Niamey, Niger, August 28, 2006.
81. Samuel Loewenberg, "Millions in Niger Facing Food Shortages Once Again," *The Lancet* 367, no. 9521 (2006): 1474–1476.
82. M. Tectonidis, "Crisis in Niger—Outpatient Care for Severe Acute Malnutrition," *The New England Journal of Medicine* 354, no. 3 (January 19, 2006): 1.
83. Ibid.
84. Interview with WFP personnel, Niamey, Niger, August 28, 2006.
85. Tectonidis, "Crisis in Niger, " 2.
86. David E. Heath, "Niger: More Than a Food Shortage," *Environment* 47, no. 10 (December 2005): 1.
87. Ibid., 1–2.
88. Ibid., 2.
89. Ibid.
90. Laurel J. Hummel, "Lowering Fertility Rates in Developing States: Security and Policy Implications for Sub-Saharan Africa," Master of Strategy Studies (MSS) thesis, United States Army War College, 2006, 13.
91. Interview with UNICEF official, Niamey, Niger, August 28, 2006.
92. K. Scharnberb, "Do-It-Yourself Famine Fight," *Chicago Tribune*, August 25, 2005, 1.
93. Climate data for figures 1.1 through 1.4 obtained from http://www.world-climate.com.
94. A. Nyong, "Drought and Conflict in the Western Sahel: Developing Conflict Management Strategies" (event summary compiled by Alison Williams of the Woodrow Wilson International Center for Scholars, October 18, 2005).
95. Simon Batterbury and Andrew Warren, "The African Sahel 25 Years After the Great Drought: Assessing Progress and Moving Towards New Agendas and Approaches." In a special issue of *Global Environmental Change* 11, no. 1 (April 2001): 1–8.
96. Mike Hulme and P. M. Kelly, "Exploring the Links Between Desertification and Climate Change," *Environment* 35, no. 6 (1993): 4–11, 39–45.
97. World Resources Institute, http://earthtrends.wri.org/text/population-health/country-profile-136.html.
98. World Bank, http://www.worldbank.org/niger.
99. J. J. McCarthy et al., *Climate Change 2001: Impacts, Adaptations and Vulnerabilities, Contribution of Working Group II to the Third Assessment Report of the Intergovernmental Panel on Climate Change*, ch. 10, 489, http://www.grida.no/climate/ipcc_tar/wg2/001.htm.

100. Hulme and Kelly, "Exploring the Links," 7–8.
101. Ibid., 7.
102. McCarthy, *Climate Change 2001*, 489–490.
103. Hulme and Kelly, "Exploring the Links," 7.
104. Ibid.
105. World Resources Institute, http://earthtrends.wri.org/text/population-health/country-profile-136.html.
106. Robert T. Watson et al., *Summary for Policymakers—the Regional Impacts of Climate Change: An Assessment of Vulnerability*, Intergovernmental Panel on Climate Change special report, November 1997, 3, http://www.ipcc.ch.
107. Ibid., 3–5.
108. Microsoft *ENCARTA*, "Niger River," 2006.
109. McCarthy, *Climate Change 2001*, 498.
110. Nyong, "Drought and Conflict," 1.
111. Interview with head of local NGO, Niamey, Niger, August 28, 2006.
112. "'Habbanae' Loans of North Niger for Harmony List," U.S. Federal News Service, June 20, 2006, 1.
113. Scharnberb, "Do-It-Yourself," 2–3.
114. Ibid., 3.
115. Taken directly from Wentling, "Draft Discussion paper; Elements of An Assistance Strategy for Niger," 2–3.
116. Ibid., 3.
117. Ibid., 32.
118. Homer-Dixon, "Environmental Scarcities and Violent Conflict," 269–271.
119. As mentioned in the introduction, this portion on Chad comes from our previous paper, with some modifications: Jebb and Abb, "Human Security and Good Governance."
120. Sachs, *The End of Poverty*, 102–103.
121. Ibid., 103.
122. Jeffrey D. Sachs, "The Geography of Economic Development," *Naval War College Review* 53, no. 4 (2000): 101–102.
123. Sachs, *The End of Poverty*, 104.
124. *Background Note: Chad*, U.S. Department of State, Bureau of African Affairs, 1. http://www.state.gov.
125. Mirjam de Bruijn and Han van Dijk, "Climate Variability and Political Insecurity: The Guera in Central Chad," unpublished paper, August 2002, 5.
126. John H. Bodley, *Anthropology and Contemporary Human Problems* (Mountain View, CA: Mayfield, 1996), 83.

127. For more discussion on the structural maladaptations introduced by global and political systems, see Bodley, *Anthropology*, 83–89.

128. de Bruijn and van Dijk, "Climate Variability," 6.

129. *Background Note: Chad*, 2.

130. Ibid., 3–5.

131. Ibid., 6.

132. Nelson Kasfir, "Sudan's Darfur: Is It Genocide?" *Current History* 104, no. 682 (May 2005): 195.

133. Ibid., 196.

134. Ibid.

135. Bureau of Democracy, Human Rights and Labor and Bureau of Intelligence and Research, "Documenting Atrocities in Darfur," Department of State Publication 11182, September 2004.

136. UNHCR, *Briefing Note on Chad*, December 6, 2004.

137. UNHCR, *Sudan Situation Update*, February 2005, 1. Note that recent reports in the fall of 2006 indicated that fighting in Sudan spilled into Chad. The town we visited, Abeche, though unharmed, reported rebel elements in the town.

138. International Crisis Group, "Islamist Terrorism in the Sahel."

139. Ibid.

140. Ibid. In our travels, the military planners we met understood the complexity of the issues affecting human survival and did not use only a counter-terrorism lens.

141. Ibid., 2.

142. Ibid., 4.

143. UNHCR, packet developed for Senator Joseph Biden's trip, May 31, 2005.

144. Interview with African Union official, Abeche, Chad, June 13, 2005.

145. UNHCR, packet for Senator Joseph Biden's trip.

146. These points were discussed with the director of the UNHCR effort and a member of the UNHCR Child Protection Office on June 13–14, 2005, in Abeche, Chad.

147. Interview with Mr. Kingsley Amaning, resident representative, UNDP, N'djamena, Chad, June 10, 2005.

148. Interview with Lt. Col. Tim Mitchell, defense attaché, Chad, N'djamena, Chad, June 10, 2005.

149. PBS, "Grand Canyon," *Nature*. Information on the program can be accessed at http://www.pbs.org/wnet/nature/grandcanyon/.

150. McCarthy, *Climate Change 2001*, 517–519.

151. Ibid., 519–520.
152. Ibid.
153. Watson et al., *Summary for Policymakers*, 4.
154. Nyong, "Drought and Conflict," 1 and Watson et al., *Summary for Policymakers*, 4.
155. Watson et al., *Summary for Policymakers*, 1–4.
156. McCarthy, *Climate Change 2001*, 519.
157. Ibid., 496–500.
158. Portions of this section related to the Horn of Africa can be found in James Forest, ed., *Training & Root Causes, The Making of a Terrorist*, vols. 2 and 3 (Westport, CT: Praeger International Security, 2005).
159. The point about speaking truth to power was based on an interview with Mr. Kingsley Amaning, resident representative of UNDP, N'djamena, Chad, June 10, 2005.
160. See Jebb and Abb, *Human Security*.
161. William Easterly, *The White Man's Burden: Why the West's Efforts to Aid the Rest Have Done So Much Ill and So Little Good* (New York: Penguin, 2006), 15–17.

CHAPTER 2

162. This chapter first appeared as Stephen F. Burgess, "Environment and Human Security in the Horn of Africa," *Journal of Human Security* 4, no. 2 (2008): 37–61 and is reprinted with the permission of the *Journal of Human Security*. Behind the chapter is an increased concern about the environment, sustainability, and security (including human security) on the part of the U.S. Department of Defense. The demand for sustainability and stabilization assessments comes from various sources within the department.
163. The views expressed are those of the author and not necessarily those of the U.S. Air War College or the Department of Defense.
164. Paul D. Williams, "Thinking About Security in Africa," *International Affairs* 83, no. 6 (2007): 1021–1038; Nana K. Poku, Neil Renwick, and Jaoa Gomes Porto, "Human Security and Development in Africa," *International Affairs* 83, no. 6 (2007): 1155–1171; Melissa Thaxton, "Integrating Population, Health and Environment in Kenya," in *Bridge: Making the Link, Population Reference Bureau* (USAID, 2007), 1–9; Kristen P. Patterson, "Integrating Population, Health and Environment in Ethiopia," in *Bridge: Making the Link, Population Reference Bureau* (USAID, 2007), 1–11.

Human security refers to security of individuals and families to have shelter, food, health, and other basic needs, as well as security from conflict, human rights abuses, environmental degradation, exploitation, disease, and other maladies. The concept of human security implies that "state security" does not guarantee security for individuals and families.

165. Michael Renner, *Fighting for Survival: Environmental Decline, Social Conflict, and the New Age of Insecurity* (New York: W.W. Norton, 1996); Thomas F. Homer-Dixon, *Environment, Scarcity, and Violence* (Princeton, NJ: Princeton University Press, 1999); Nils Petter Gleditsch, "Armed Conflict and the Environment: A Critique of the Literature," *Journal of Peace Research* 35, no. 3:381–400.

166. Sanjeev Khagram, William C. Clark, and Dana Firas Raad, "From the Environment and Human Security to Sustainable Security and Development," *Journal of Human Development* 4, no. 2 (2003): 289–313.

167. See *Mapping Climate Vulnerability and Poverty in Africa: Where Are the Hot Spots of Climate Change and Household Vulnerability? Report to the Department for International Development*, submitted by the International Livestock Research Institute (ILRI), Nairobi, Kenya, in collaboration with the Energy and Resources Institute (TERI), New Delhi, India, and the African Centre for Technology Studies (ACTS), Nairobi, Kenya, May 2006, http://www.acts.or.ke/ pubs/books/docs/Mapping_Vuln_Africa.pdf. See also Stephan Faris, "The Real Roots of Darfur," *Atlantic Monthly*, April 2007. Faris points to global warming and resulting desertification as major causes of the conflict between Janjaweed pastoralists and Darfur farmers.

168. Peter Chalk, "Case Study: The East African Corridor," in *Ungoverned Territories: Understanding and Reducing Terrorist Risks*, ed. Angel Rabasa, Steven Boraz, Peter Chalk, Kim Cragin, Theodore W. Karasik, Jennifer D. P. Moroney, Kevin A. O'Brien, and John E. Peterset (Santa Monica, CA: RAND Project AIR FORCE, MG-561-AF, 2007), 147–172. The report is also available at http://www.rnd.org/pubs/mongraphs/MG561/.

169. USAID, *A Better Way of Working: Creating Synergies in the Pastoral Zones of the Greater Horn of Africa* (Nairobi, Kenya, Regional Enhanced Livelihoods in Pastoral Areas (RELPA): RFA—Support Documents, November 2001).

170. CJTF-HOA, www.hoa.centcom.mil; interview with Major General Timothy Ghormley, CENTCOM deputy J-3, Tampa, Florida, May 14, 2007. Maj. Gen. Ghormley was CJTF-HOA commander in 2006.

171. Field research undertaken by the author in May and June 2007 in Ethiopia and Kenya focused on these issues. See Ginny Hill, "Military Focuses

on Development in Africa: In Djibouti, US Forces Combat Terrorism with Civil Affairs Work. Will This Be a Model for a Future US Military Command in Africa?" *Christian Science Monitor*, June 22, 2007. See also Combined Joint Task Force—Horn of Africa, "Mission and Philosophy," http://www.hoa.centcom.mil/resources/english/facts.asp.

172. Hans Günter Brauch, ed., *Security and Environment in the Mediterranean: Conceptualising Security* (Berlin: Springer, 2003). The volume draws links between environmental degradation and human insecurity.

173. Roland Paris, "Human Security—Paradigm Shift or Hot Air?" *International Security* 26, no. 2 (2001): 87–102.

174. Williams, "Thinking about Security in Africa," 1021–1038; Poku, Renwick, and Porto, "Human Security and Development in Africa," 1155–1171.

175. Desertification is known in international development circles as "land degradation in the drylands."

176. Oli Brown, Anne Hammill, and Robert McLeman, "Climate Change as the 'New' Security Threat: Implications for Africa," *International Affairs* 83, no. 6 (2007): 1141–1154. Global warming is the principal form of climate change.

177. *Mapping Climate Vulnerability and Poverty in Africa.*

178. Faris, "The Real Roots of Darfur."

179. United Nations Fund for Population Activities (UNFPA), *Global Population and Water, 2003*, http://www.unfpa.org/publications/detail.cfm?ID= 68&filterListType. UNFPA, *State of the World Population 2004*, http://www.unfpa.org/swp/2004/pdf/summary.pdf.

180. M. V. K. Sivakumar, "Interactions between Climate and Desertification," *Agriculture and Forest Methodology* 142, nos. 2–4 (February 2007): 143–155.

181. Simon A. Mason, Tobias Hagmann, Christine Bichsel, Eva Ludi, and Yacob Arsano, "Linkages between Sub-national and International Water Conflicts: The Eastern Nile Basin," in *Facing Global Environmental Change: Environmental, Human, Energy, Food, Health and Water Security Concepts*, ed. Hans Günter Brauch, John Grin, Czelaw Mesasz, Heinz Krummenacher, Navnita Chadha Behera, Bechir Chourou, Ursula Oswald-Spring, P. H. Liotta, and Patricia Kameri-Mbote (Berlin: Springer-Verlag, 2007), 325–334.

182. Bruce Lankford and Thomas Beale, "Equilibrium and Non-equilibrium Theories of Sustainable Water Resources Management: Dynamic River Basin and Irrigation Behaviour in Tanzania," *Global Environmental Change* 17, no. 2 (May 2007): 168–180. Rivers are a source of irrigation, fish, and navigation.

183. UNFPA, *Global Population and Water*. Microlevel water shortages include a deficit of boreholes and falling water tables.
184. Peter Kagwanja, "Calming the Waters: The East African Community and Conflict over the Nile Resources," *Journal of Eastern African Studies* 1, no. 3 (November 2007): 321–337.
185. Cormac Ó Grada, "Making Famine History," *Journal of Economic Literature* 45 (March 2007): 3–36.
186. Ibid. The reliance on genetically modified seeds has contributed to famine in some cases.
187. Renner, *Fighting for Survival*, 17–30.
188. Gleditsch, "Armed Conflict."
189. Homer-Dixon, *Environmental Scarcity and Violence*, 14–15.
190. Ibid.
191. Ibid., 6.
192. Ibid., 137–141. Freshwater is one of the four resources that Homer-Dixon identifies as a likely spark for conflict. The other three are agriculturally productive land, forests, and fish.
193. Ibid., 139.
194. John Waterbury, *The Nile Basin: National Determinants of Collective Action* (New Haven, CT: Yale University Press, 2001), 28.
195. Peter Gleick, "Water Conflict Chronology," Pacific Institute for Studies in Development, Environment, and Security, http://www.worldwater.org/conflict.htm. There have been only twenty-four interstate disputes resulting in the use or threatened use of military force in recorded history from political, military, or developmental issues over water between two or more states.
196. Magnus Ole Theisen and Kristian Bjarnøe Brandsegg, "The Environment and Non-state Conflicts in Sub-Saharan Africa" (paper presented at International Studies Association convention, March 1–3, 2007). The study focuses on so-called "nonstate" conflicts for sub-Saharan Africa in the period 2002–2005 and also finds that higher levels of rainfall coincide with a higher risk of conflict onset. This paper reflects recent advances in research on environmental scarcity and conflict using geospatial analysis.
197. Interview with Mebrate Woldentensaie, research and training officer, the Conflict and Early Warning Response Mechanism (CEWARN) of the Inter-Governmental Authority on Development (IGAD), Addis Ababa, Ethiopia, May 31, 2007.
198. Stephen F. Burgess, "African Security in the 21st Century: The Challenges of Indigenization and Multilateralism," *African Studies Review* 41, no. 2 (September 1998): 37–61.

199. Intergovernmental Authority on Development, "History of IGAD," http://www.igad.org/about/index.html. In the 1990s, IGAD added a peacemaking role and helped end conflicts in southern Sudan and Somalia. Eritrea joined IGADD in 1993 at independence and withdrew from IGAD in 2006 for political reasons (differences with Ethiopia over Somalia).

200. IGAD, "IGAD Strategy, October 2003," http://www.igad.org/about/igad_strategy.pdf. IGAD has not lived up to expectations as a regional organization that would foster regional cooperation.

201. *Mapping Climate Vulnerability and Poverty in Africa*; Faris, "The Real Roots of Darfur." Floods in Uganda, Somalia, and other countries in September 2007 seemed to be indicative of the effects of global warming.

202. Interview with Joan Kariuki, Andrew Adwerah, and Elvin Nyukuri, Africa Centre for Technology Studies (ACTS), World Agroforestry Centre (ICRAF), Nairobi, Kenya, June 7, 2007.

203. Interview with Philip Dobie, director, UN Development Programme Drylands Development Center, Nairobi, Kenya, June 6, 2007.

204. United Nations Environment Program, *Sudan: Post-Conflict Environmental Assessment, June 2007*, http://www.unep.org/sudan/.

205. Ibid.

206. Ibid.

207. Ibid. Deforestation was over 0.84 percent per annum at the national level and 1.87 percent per annum in UNEP case-study areas.

208. Ibid.

209. Interview with Walter Knausenberger, senior regional environmental officer, USAID, Nairobi, Kenya, June 5, 2007.

210. Interview with Philip Dobie, director, UN Development Programme Drylands Development Center, Nairobi, Kenya, June 6, 2007. The UNDP Drylands Agency is dedicated to improving the way of life of pastoralists.

211. M. A. Mohamed Salih, introduction to *African Pastoralism: Conflict, Institutions and Government*, ed. M. A. Mohamed Salih, Ton Dietz, and Abdel Mohamed Ahmed, 1–22 (London: Pluto Press, 2001). See also Paul Spencer, *The Pastoral Continuum: The Marginalization of Tradition in East Africa* (Oxford: Oxford University Press, 1998).

212. UN Office for the Coordination of Humanitarian Affairs-Pastoral Communication Initiative (UN OCHA-PCI) and Institute of Development Studies (UK), "Is Pastoralism Still Viable in the Horn of Africa? New Perspectives from Ethiopia, Addis Ababa: UN OCHA-PCI and IDS-UK Report May, 2006. U.S. Agency for International Development. Household Livestock Holding & Livelihood Vulnerability Status Report on PRA

and Herd Structure Study," in Dollo Ado Woreda, *Libane Zone of Somali National Regional State*, ed. Dollo Ado Woreda, 4–5 (March 2006).

213. U.S. Agency for Internatinal Development, Household Livestock Holding & Livelihood Vulnerability Status Report on PRA and Herd Structure Study," in *Libane Zone of Somali National Regional State*, ed. Dollo Ado Woreda, 4–5 (March 2006).

214. Interview with Philip Dobie, director, UN Development Programme Drylands Development Center, Nairobi, Kenya, June 6, 2007.

215. John K. Livingstone and A. Livingstone, *A Comparative Study of Pastoralist Parliamentary Groups: Kenya Case Study*, NRI/PENHA Research Project on Pastoralist Parliamentary Groups, funded by the British Department for International Development's (DFID's) Livestock Production Programme and the CAPE Unit, African Union's Interafrican Bureau of Animal Resources (AU-IBAR, May 2005), http://www.nri.org/projects/pastoralism/kenyappgmayfinal.pdf.

216. Interview with Mebrate Woldentensaie, research and training officer, the Conflict and Early Warning Response Mechanism (CEWARN) of the Inter-Governmental Authority on Development (IGAD), Addis Ababa, Ethiopia, May 31, 2007.

217. Interview with John Graham and Yacob Wondimkum, U.S. Agency for International Development, Addis Ababa, Ethiopia, May 31, 2007.

218. Ibid. In Somali areas with small clans, there is no conflict. However, in Gambella is another area where USAID has been assisting with conflict-mitigation work, even though environmental issues are not relevant in such a lush and sparsely populated area.

219. Interview with Simon Richards, "Pact" director for African Region, June 5, 2007.

220. Interview with Philip Dobie, director, UN Development Programme Drylands Development Center, Nairobi, Kenya, June 6, 2007.

221. Ibid.

222. Interview with Walter Knausenberger, senior regional environmental officer, USAID, East Africa Regional Mission, Nairobi, Kenya, June 5, 2007.

223. Interview with Philip Dobie, director, UN Development Programme Drylands Development Center, Nairobi, Kenya, June 6, 2007.

224. Interview with Walter Knausenberger, senior regional environmental officer, USAID, East Africa Regional Mission, Nairobi, Kenya, June 5, 2007.

225. Interview with Tewolde Ezgabier, Ethiopian Environmental Protection Agency, Addis Ababa, Ethiopia, May 29, 2007. Terraces that existed for

two thousand years have deteriorated only in the last century. Huge gullies have formed due to the failure of the land management system.

226. Patrick Webb and Joachim von Braun, *Famine and Food Security in Ethiopia: Lessons for Africa* (New York: Wiley, 1994).

227. Interview with John Graham and Yacob Wondimkum, U.S. Agency for International Development, Addis Ababa, Ethiopia, May 31, 2007.

228. Ibid. The government can control the population easier in the rural areas than in the urban areas, where the people tend to support opposition political parties.

229. Interview with Mekri Argaw, professor, School of Natural Science, Addis Ababa University, Addis Ababa, Ethiopia, May 29, 2007.

230. Thaxton, "Integrating Population, Health and Environment in Kenya," 1–9. Patterson, "Integrating Population, Health and Environment in Ethiopia," 1–11.

231. U.S. Department of Energy, Energy Information Administration, "Environment and Renewable Energy in Africa," *Energy in Africa*, http://www.eia.doe.gov/emeu/cabs/Archives/africa/chapter7.html. Interview with Lisa Brodey, U.S. Embassy, Addis Ababa, Ethiopia, May 29, 2007.

232. Interview with Lisa Brodey, U.S. Embassy, Addis Ababa, Ethiopia, May 29, 2007.

233. UNFPA, *State of the World Population 2004*; Thaxton, "Integrating Population, Health and Environment in Kenya," 1–9; Patterson, "Integrating Population, Health and Environment in Ethiopia," 1–11.

234. Interview with Lisa Brodey, U.S. Embassy, Addis Ababa, Ethiopia, May 29, 2007.

235. Extension services and environmental management agencies are especially relevant indicators of state capacity. Markets and credit for farmers and pastoralists are especially relevant indicators of private-sector development.

236. An example of such interventions has been the much-criticized Counterterrorism Joint Task Force-Horn of Africa (CJTF-HOA) model for overcoming environmental degradation and conflict and extremism.

237. Interview with Joan Kariuki, Andrew Adwerah, and Elvin Nyukuri, Africa Centre for Technology Studies (ACTS), World Agroforestry Centre (ICRAF), Nairobi, Kenya, June 7, 2007. According to Joan Kariuki and her colleagues at the African Centre for Technology Studies, tree planting is one of the best ways to mitigate environmental degradation. A good example is Wangari Maathai and the Greenbelt Movement.

238. Interview with Walter Knausenberger, senior regional environmental officer, USAID, East Africa Regional Mission, Nairobi, Kenya, June 5, 2007.

239. UNEP, *Sudan: Post-Conflict Environmental Assessment*.

240. Ibid.

241. Interview with Ann Muir, Customary Pastoral Institutions Study, USAID, Nairobi, Kenya, March 2007.

242. Andy Catley, ed., *Livestock and Pastoral Livelihoods in Ethiopia: Impact Assessments of Livelihoods-Based Drought Interventions in Moyale and Dire Woredas*, Pastoralist Livelihood Initiative, USAID and Feinstein International Center, Friedman School of Nutrition Science and Policy, Tufts University, Addis Ababa, Ethiopia, 2007.

243. Interview with Walter Knausenberger, senior regional environmental officer, USAID, East Africa Regional Mission, Nairobi, Kenya, June 5, 2007. USAID has started a donor group focused on climate change and conducted seven to nine province assessments in Kenya with a coordinated policy approach, which has resulted in better potential for coverage of affected areas.

244. Interview with Philip Dobie, director, UN Development Programme Drylands Development Center, Nairobi, Kenya, June 6, 2007.

245. Ibid.

246. Interview with Paul McDermott, office director, Regional Conflict Management and Governance Office, Regional Economic Development Services Office, East and Southern Africa, USAID, Nairobi, Kenya, June 7, 2007.

247. Interview with Walter Knausenberger, senior regional environmental officer, USAID, East Africa Regional Mission, Nairobi, Kenya, June 5, 2007.

248. Ibid.

249. Interview with Tewolde Ezgabier, Ethiopian Environmental Protection Agency, in Addis Ababa, Ethiopia, May 29, 2007.

250. Interview with John Graham and Yacob Wondimkum, U.S. Agency for International Development, Addis Ababa, Ethiopia, May 31, 2007.

251. Ibid. The land registration process is not having a serious gender impact in the Christian highlands, but in Muslim areas, it is.

252. Interview with Dr. Mulugeta Feseha, director, Institute of Development Research, Addis Ababa University, Addis Ababa, Ethiopia, May 30, 2007.

253. Interview with John Graham and Yacob Wondimkum, U.S. Agency for International Development, Addis Ababa, Ethiopia, May 31, 2007.

FAO tried planting mesquite trees, without success, because they prevented the reestablishment of natural grass, negatively affected acacia trees, and left no fodder. Eucalyptus trees deplete groundwater.

254. Interview with Tewolde Ezgabier, Ethiopian Environmental Protection Agency, Addis Ababa, Ethiopia, May 29, 2007.

255. Ibid.

256. Ibid.

257. Ibid.

258. Interview with Philip Dobie, director, UN Development Programme Drylands Development Center, Nairobi, Kenya, June 6, 2007. According to Dobie, governments must ask if they want UNDP to play a lead coordinating role.

259. USAID, *A Better Way of Working: Creating Synergies in the Pastoral Zones of the Greater Horn of Africa, Nairobi, Kenya, Regional Enhanced Livelihoods in Pastoral Areas (RELPA): RFA—Support Documents*, November 2001.

260. For example, USAID, *A Better Way of Working (RELPA)*.

261. Interview with Simon Richards, "Pact" director for African Region, Nairobi, Kenya, June 5, 2007.

262. Ibid. In discussion is a broader regional conflict-mitigation program. Security is needed for NGOs specializing in conflict mitigation and doing work in conflict zones.

263. Interview with Walter Knausenberger, senior regional environmental officer, USAID, East Africa Regional Mission, Nairobi, Kenya, June 5, 2007.

264. USAID, *Mandera-Gedo Cross-Border Conflict Mitigation Initiative Assessment Team Report*, March 2006.

265. Interview with Mebrate Woldentensaie, research and training officer, the Conflict and Early Warning Response Mechanism (CEWARN) of the Inter-Governmental Authority on Development (IGAD), Addis Ababa, Ethiopia, May 31, 2007.

266. Ibid.

267. Interview with Paul McDermott, office director, Regional Conflict Management and Governance Office, Regional Economic Development Services Office, East and Southern Africa, USAID, Nairobi, Kenya, June 7, 2007.

268. Interview with Major General Timothy Ghormley, CENTCOM deputy J-3, Tampa, Florida, May 14, 2007. Maj. Gen. Ghormley was CJTF-HOA commander in 2006.

269. Ghormley diagrammed the CJTF-HOA counterinsurgency and counterter-rorism model, which was devised by him and his predecessor, Maj. Gen. Sam Helland, who both were associated with the Civil Operations and Rural Development Support Program (CORDS) and the "hearts and minds" campaigns of the 1960s in Vietnam. Maj. Gen. Ghormley recommended the CJTF-HOA model be adopted by Africa Command and expanded to north, west, central, and southern Africa. "Command Assist Support Teams" (CASTs) would be established (similar to provincial reconstruction teams), working with coalition forces, international and national agencies, and NGOs. The CASTs would perform tasks that would fulfill U.S. embassies' Mission Performance Plans (MPPs). For example, CASTs could develop water projects, which are part of every embassy's MPPs. The CASTs could help develop "Regional Centers for Excellence" where the locals would follow a syllabus in water project management. Maj. Gen. Ghormley asserted that the Combined Joint Task Force-Horn of Africa (CJTF-HOA) serves as a model for Africa Command (AFRICOM) (and CENTCOM) in winning hearts and minds in the war on terror. A model of the "coalition of the willing to act" would mean that the United States would no longer act as the intermediary between African states, European states, China, Japan, and India. For example, this would mean that the coalition engages the Africans in water projects, with the Chinese, Japanese, or Germans in charge. Maj. Gen. Ghormley recommended that AFRICOM work on a broad base and enable Africans to take charge of African security. The broader base should include a coalition of those willing to act, including the European Union (EU), Russia, China, and India. This coalition should work on a broad range of environmental and human security issues, including water and disease. For example, avian flu could kill 30 million Africans in the first wave, which would require a coalition to manage.

270. Interview with Major General Timothy Ghormley, CENTCOM deputy J-3, Tampa, Florida, May 14, 2007.

271. For example, the Gambella, Dinka, and Nuer.

272. Interview with Commander Ken Atkins, CJTF-HOA, U.S. Embassy, Nairobi, Kenya, June 8, 2007.

273. Interview with Major Errill Avecilla, civil affairs liaison officer, and Commander Ken Atkins, CJTF-HOA, U.S. Embassy, Nairobi, Kenya, June 8, 2007.

274. Interview with Major Errill Avecilla, civil affairs liaison officer, and Commander Ken Atkins, CJTF-HOA, U.S. Embassy, Nairobi, Kenya, June 8, 2007.

275. Interview with Lt. Col. Scott Rutherford, defense attaché, U.S. Embassy, Nairobi, Kenya, June 4, 2007.
276. Interview with Walter Knausenberger, senior regional environmental officer, USAID, East Africa Regional Mission, Nairobi, Kenya, June 5, 2007.
277. Interview with Eric Wong, U.S. Embassy, Ethiopia, May 30, 2007.
278. Interview with Craig White, deputy political counselor, U.S. Embassy, Nairobi, Kenya, June 8, 2007.
279. Interview with Philip Dobie, director, UN Development Programme Drylands Development Center, Nairobi, Kenya, June 6, 2007.
280. Interview with Simon Richards, "Pact" director for African Region, Nairobi, Kenya, June 5, 2007.
281. Alan Nicol, "The Dynamics of River Basin Cooperation: The Nile and Okavango basins," in *Transboundary Rivers, Sovereignty and Development: Hydropolitical Drivers in the Okavango River Basin*, ed. Anthony Turton, Peter Ashton, and Eugene Cloete, 167–186 (Pretoria: University of Pretoria, 2003).
282. Eritrea participates as an observer nation in the Nile Basin Initiative.
283. Blue Nile issues are covered by the Eastern Nile Subsidiary Action Program (ENSAP) and the Eastern Nile Technical Regional Office (ENTRO), based in Addis Ababa, Ethiopia.
284. Nile Basin Initiative, http://www.nilebasin.org/nbibackground.htm.
285. Nile Basin Initiative, http://www.nilebasin.org/SVP_Overview.htm.
286. Ashok Swain, "The Nile River Basin Initiative: Too Many Cooks, Too Little Broth," *SAIS Review* 22, no. 2 (summer-fall 2002): 303.
287. Anthony Turton, GIBB-SERA chair in integrated water resource management, Environmentek, CSIR, Pretoria, South Africa, e-mail message to the author, August 5, 2005.
288. Swain, "The Nile River Basin Initiative," 307.
289. Waterbury, *The Nile Basin*, 137–139.
290. Interview with Johnson Akinbola Oguntola, senior regional adviser on integrated water resources management, Sustainable Development Division, UN Economic Commission for Africa, Addis Ababa, Ethiopia, May 29, 2007.
291. Edwin Musoni, "Africa: Rift Widens Over Nile Basin Pact as Egypt, Sudan Remain Reluctant," *The New Times* (Kigali), February 29, 2008.
292. Benson Kathuri, "Egypt No Longer Has the Power to Dictate over Waters," *East African Standard*, February 23, 2005, 1.
293. Joyce Mulama, "The Fate of the Nile in the Spotlight," Inter Press Service News Agency, March 16, 2004, 1.

294. Global Water Partnership East Africa, *Partnership for Africa's Water Development 1 Program* (PAWD), Q1 Progress Report submitted to the Global Water Partnership Organization (Stockholm, Sweden, 2005), 11–19, http://www.gwpena.org/dwn/PAWD%201_2005Annual%20report%20_GWP%20EnA.pdf.

295. Interview with Lisa Brodey, Regional Environmental Program, U.S. Embassy, Addis Ababa, Ethiopia, May 29, 2007.

296. Interview with Kinfe Abraham, president of the Ethiopian International Institute for Peace and Development, Addis Ababa, Ethiopia, May 31, 2007. See also Kinfe Abraham, "No Treaty to Prevent Ethiopia from Utilizing the Nile Water," *Addis Tribune*, February 25, 2005, http://www.addistribune.com/Archives/2005/02/25-02-05/No.htm. Abraham is president of the Ethiopian International Institute for Peace and Development.

297. Interview with Tewolde Ezgabier, Ethiopian EPA, Addis Ababa, Ethiopia, May 29, 2007.

298. Robert O. Collins, *The Nile* (New Haven, CT: Yale University Press, 2002), 228.

299. Swain, "The Nile River Basin Initiative," 299.

300. Ibid., 297. See also African Development Bank, *Policy for Integrated Water Resources Management, April 2000*, http://www.afdb.org/pls/portal/docs/PAGE/ADB_PAGEGROUP/TOPICS/ADBPROJECTSEC-TORS/WATERSUPPLYANDSANITATION/INTEGRATED_WATER_POLICY_APR2000.PDF.

301. Roger Thurow, "Unable to Tap Power of the Nile, Ethiopia Relies on Fuel Carriers," *The Wall Street Journal*, November 26, 2003, A8.

302. Swain, "The Nile River Basin Initiative," 301.

303. David Read Barker, (address to Lake Tana Symposium, Bahir Dar University, Bahir Dar, Ethiopia, September 24, 2004), http://www.world-lakes.org/uploads/Lake_Tana_ Symposium24Sep04_drb.pdf.

304. Interview with Lisa Brodey, U.S. Embassy, Addis Ababa, Ethiopia, May 29, 2007. Interview with Yacob Arsano, dean, College of Social Sciences, Addis Ababa University, Addis Ababa, Ethiopia, May 30, 2007.

305. Waterbury, *The Nile Basin*, 128–130.

306. Interview with Lisa Brodey, U.S. Embassy, Addis Ababa, Ethiopia, May 29, 2007.

CHAPTER 3

307. The views expressed are those of the author and not necessarily those of the U.S. Marine Corps, the U.S. Naval Academy, or the Department of Defense.

308. For an excellent discussion of the divisions within the Jola ethnic group, see Louis Vincent Thomas, *Les Diola: Essai D'Analyse Fonctionnelle sur une Population de Basse-Casamance* (Amsterdam: Swets and Zeitlinger, 1978). Thomas identifies nine different ethnic groups within the "Jola nation." Ethnographers have looked at the Jola from linguistic, historical, and geographical perspectives. Their findings show that the Jola were relatively recent (circa 1900) arrivals to the Lower Casamance, where the majority of the conflict has occurred. In fact, the original autochthones of the Casamance were not Jola but, rather, Bainouk.

309. Interview with Robert Lloyd, Pepperdine University, in Montgomery, Alabama, at a conference on African security cosponsored by the Association for the Study of the Middle East and Africa and the U.S. Air Force Research Institute, February 13, 2009.

310. Mark Doyle, "Senegal's Southern Rebellion Starts to End," BBC News, December 16, 2004, http://news.bbc.co.uk/go/pr/fr/-/2/hi/Africa/4102267. stm.

311. BBC News, "Crowds Cheer Senegal Peace Deal," December 30, 2004, http://news.bbc.co.uk/go/pr/fr/-/2/hi/africa/4135797.stm.

312. Michael C. Lambert, "Violence and the War of Words: Ethnicity v. Nationalism in the Casamance," *Africa: Journal of the International African Institute* 68, no. 4 (1998): 592.

313. Doyle, "Senegal's Southern Rebellion Starts to End."

314. UN Integrated Regional Information Networks (IRIN), "Senegal: Finding Incentives for Peace in Casamance," AllAfrica.com, June 25, 2008, http://allafrica.com/stories/200806251043.html

315. Ferdinand de Jong and Geneviève Gasser, "Contested Casamance/Discordante Casamance," *Canadian Journal of African Studies/Revue Canadienne des Études Africaines* 39, no. 2 (2005): 213.

316. Some scholars spell "Jola" as "Diola." Others spell it "Joola." I spell it "Jola" in all cases, regardless of the source's spelling, in the interest of standardization for this chapter.

317. Primarily Wolof, but perhaps some Fulani and Toucouleur as well.

318. Numerous personal interviews with various government and civil society actors in the Casamance, 2005–2007.

319. For further discussion, see Robert I. Rotberg, "The New Nature of Nation-State Failure," *The Washington Quarterly* 25, no. 3 (summer 2002): 87–89.

320. Sambou had to resign shortly after the September 2002 Joola ferry disaster, in which around two thousand people died when the Dakar-Ziguinchor ferry capsized off the coast of the Gambia. For further information, see BBC News, "Ministers Quit over Senegal tragedy" October 2, 2002, http://news.bbc.co.uk/2/hi/africa/2290879.stm.

321. Lambert, "Violence and the War of Words," 591.

322. Ibid.

323. Hamadou Tidiane Sy, "The Casamance Separatist Conflict: From Identity to the Trap of 'Identitism,'" in *Identity Matters: Ethnic and Sectarian Conflict*, ed. James L. Peacock, Patricia M. Thornton, and Patrick B. Inman, 159–160 (Oxford: Berghahn Books, 2007).

324. Mamadou Diouf, "Between Ethnic Memories and Colonial History in Senegal: The MFDC and the Struggle for Independence in Casamance," in *Ethnicity and Democracy in Africa*, ed. Bruce Berman, Dickson Eyoh, and Will Kymlicka, 223 (Oxford: Oxford University Press, 2004).

325. *London Times*, August 7, 1890, as quoted in Robert Tignor, Jeremy Adelman, Stephen Aron, Peter Brown, Benjamin Elman, Stephen Kotkin, Xinru Liu, Suzanne Marchand, Holly Pittman, Gyan Prakash, Brent Shaw, and Michael Tsin, *Worlds Together, Worlds Apart: A History of the World from the Beginnings of Humankind to the Present*, vol. C., 2nd ed. (New York: W.W. Norton and Company, 2008), 744.

326. John Milner Gray, *A History of the Gambia* (London: Frank Cass, 1966), 464.

327. *Annual Report for 1849*, as quoted in Gray, *A History of the Gambia*, 458.

328. Letter from Father Augustin Diamacoune Senghor to President Abdou Diouf, May 12, 1982, as quoted in Diouf, "Between Ethnic Memories and Colonial History in Senegal," 223.

329. Sy, "The Casamance Separatist Conflict," 166.

330. Lambert, "Violence and the War of Words," 589.

331. Sud Hebdo, no. 90 (February 1, 1990), as quoted in Diouf, "Between Ethnic Memories and Colonial History in Senegal," 235.

332. Lambert, "Violence and the War of Words," 590.

333. *Archives Nationales de la République du Sénégal* (1944), 11, in Lambert, "Violence and the War of Words," 590.

334. The Wolof, who are almost 100 percent Muslim, are the largest ethnic group in Senegal, making the Wolof language the most widely spoken indigenous language in Senegal and the Gambia.

335. Lambert, "Violence and the War of Words," 590.

336. Author's translation of Jordi Tomàs, "La Parole de Paix n'a Jamais Tort: La Paix et la Tradition dans le Royaume d'Oussouye," *Canadian Journal of African Studies/Revue Canadienne des Etudes Africaines* 39, no. 2 Contested Casamance/Discordante Casamance (2005): 426.

337. Diouf, "Between Ethnic Memories and Colonial History in Senegal," 236.

338. UN Integrated Regional Information Networks, "Senegal: Finding Incentives for Peace in Casamance."

339. Interview with local businessman in Guinea-Bissau, March 2009. Gérard Prunier corroborates this information in his discussion of the roots of the Congo conflict spawned by the 1994 Rwandan genocide in *Africa's World War: Congo, the Rwandan Genocide, and the Making of A Continental Catastrophe* (Oxford: Oxford University Press, 2009), 27.

340. Renegade troops of the armed forces of Guinea-Bissau assassinated President Vieira on March 2, 2009, following the assassination of their armed forces chief of staff, Major General Tagme Na-waie, who was known as a warrior's warrior and a man of the troops.

341. In much the same fashion, Haitian slaves, on the eve of their rebellion against the French in 1791, appealed to their ancestors in a similar traditional African ceremony in another "sacred forest." Tignor et al. write: "At a secret forest meeting held on August 14, 1791, the persons who were to lead the initial stage of the revolution gathered to affirm their commitment to one another at a voodoo ritual, presided over by a tall, black priestess, 'with strange eyes and bristly hair'" (*Worlds Together, Worlds Apart,* 653). Regardless of the degree of penetration by Christianity and Islam, traditional animist beliefs are prevalent throughout Africa.

342. There seems to be some progress on this front. According to an article in *Sud Quotidien,* an independent daily Senegalese newspaper, on February 26, 2009, U.S. Ambassador Marcia Bernicat met with a traditional Jola king, "Man" Majesty Diola, in the village of Oussouye (Father Diamacoune's home village) during "a working visit" in February 2009.

343. Interview with a former Senegalese government minister who had experience in the Sacred Forest, 2006.

344. A former U.S. military attaché to Senegal scoffed that the Casamance conflict was a "pseudoconflict" and not worthy of concern as a real civil war like those found in other parts of Africa.

345. UN IRIN, "Senegal: Finding Incentives for Peace in Casamance."

346. UN Office on Drugs and Crime, *Drug Trafficking As A Security Threat in West Africa,* November 2008, 5.

347. UN IRIN, "Senegal: Finding Incentives for Peace in Casamance." For a discussion of how aid has perpetuated conflict in other parts of Africa, see John Prendergast, *Frontline Diplomacy: Humanitarian Aid and Conflict in Africa* (Boulder, CO: Lynne Rienner, 1996).
348. U.S. military affairs in every African country except Egypt became the responsibility of U.S. Africa Command (AFRICOM) on November 1, 2008.
349. International Campaign to End Landmines, *Landmine Monitor Report, Guinea-Bissau, 2006*, July 1, 2006, http://www.icbl.org/lm/country/guinea_bissau/.
350. The Trans-Sahara Counterterrorism Partnership consisted of, besides the United States, Algeria, Morocco, Tunisia, Mauritania, Mali, Niger, Chad, Senegal, and Nigeria.
351. Emily Hunt, "Al-Qaeda's North African Franchise: The GSPC Regional Threat," Washington Institute for Near East Policy, Policy Watch no. 1034, September 28, 2005, http://www.washingtoninstitute.org/templateC05.php?CID=2379.
352. USAID Office of Inspector General, "Audit of USAID/Senegal's Casamance Conflict Resolution Program," Report No. 7-685-03-003-P, May 30, 2003, 6.
353. UN IRIN, "Senegal: Neither War Nor Peace After 25 Years," December 27, 2007, http://www.irinnews.org/Report.aspx?ReportId=76015.

Chapter 4

354. Turton, *Three Strategic Water Quality Challenges*.
355. Ibid.
356. Anthony Turton, "South African Water and Mining Policy: A Study of Strategies for Transition," in *Water Transitions*, ed. D. Huitema and S. Meijerink (The Netherlands: Edgar Elger, 2009).
357. T. R. H. Davenport and C. Saunders, *South Africa: A Modern History* (Houndsmills, UK, and New York: Macmillan Press and St. Martin's Press, 2000); M. M. Evans, *The Boer War: South Africa 1899–1902, Osprey Military* (Mechanicsburg, PA: Stackpole Books, 1999); Greg Mills and David Williams, *Seven Battles that Shaped South Africa* (Cape Town: Tafelberg, 2006).
358. R. Edgecombe, "The Mfecane or Difaqane," in *New History of South Africa in Word and Image* (Afrikaans), ed. T. Cameron and S. B. Spies (Cape Town: Human and Rousseau, 1986); Frank Welsh, *A History of South Africa* (London: Harper Collins Publishers, 2000).

359. Welsh, *A History of South Africa*, 140.
360. Martin Meredith, *Diamonds, Gold and War: The Making of South Africa* (Johannesburg: Jonathan Ball, 2007), 366.
361. Ibid.
362. Ibid.
363. L. M. Phillips, *With Rimington in the Boer War* (London: Edward Arnold, 1901); Meredith *Diamonds*.
364. Welsh, *A History of South Africa*, 343.
365. Meredith, *Diamonds*, 457.
366. E. Hobhouse, *1901 Report of a Visit to the Camps of Women and Children in the Cape and Orange River Colonies* (London: Friars Printing Association, 1907).
367. Marouf Hasian, "The 'Hysterical' Emily Hobhouse and Boer War Concentration Camp Controversy," *Western Journal of Communication* (March 2003), http://www.accessmylibrary.com/coms2/summary_0286-23546431_ITM.
368. M. H. Fawcett, *The Concentration Camps in South Africa* (London: Westminster Gazette, 1901).
369. Parliamentary Debates, "South African War Mortality in the Camps of Detention. 1901, June 17," in *The Parliamentary Debates, 4th series, 2nd sess. of the 27th Parliament of the United Kingdom of Great Britain and Ireland*, vol. 95 (London: Wyman and Sons, 1901); S. B. Spies, *Methods of Barbarism: Roberts, Kitchener and Civilians in the Boer Republics January 1900–May 1902* (Cape Town: Human and Rousseau, 1977).
370. Meredith, *Diamonds*.
371. W. Gutteridge, "Mineral Resources and National Security," *Conflict Studies* 162 (1984): 1–25. Reprinted in *South Africa: From Apartheid to National Unity, 1981–1994*, ed. W. Gutteridge, 59–83 (Aldershot, Hants, UK, and Brookfield, VT: Dartmouth Publishing, 1995).
372. V. Percival and T. Homer-Dixon, *Environmental Scarcity and Violent Conflict: The Case of South Africa* (Washington, DC: American Association for the Advancement of Science, 1995); V. Percival and T. Homer-Dixon, "Environmental Scarcity and Violent Conflict: The Case of South Africa," *Journal of Peace Research* 35, no. 3 (1998): 279–298; V. Percival and T. Homer-Dixon, "The Case of South Africa," in *Environmental Conflict*, ed. P. F. Diehl and N. P. Gleditsch, 13–35 (Boulder, CO: Westview Press, 2001); Anthony Turton, C. Schultz, H. Buckle, M. Kgomongoe, T. Malungani, and M. Drackner, "Gold, Scorched Earth and Water: The Hydropolitics of Johannesburg," *Water Resources Development* 22, no. 2 (2006): 313–335.

373. Turton, "South African Water."

374. S. Bega, "Water Expert Suspended," Environment South Africa, http://www.environment.co.za/topic.asp?TOPIC_ID=2179; Carol Paton, "Dam Dirty," *Financial Mail*, November 2008, 32–39; M. Power, "CSIR Suspends Top Water Scientist," *The Times*, November 23, 2008, http://www.thetimes.co.za/PrintEdition/Insight/Article.aspx?id=890387; Anthony Turton, *Three Strategic Water Quality Challenges*.

375. Anthony Turton, "Precipitation, People, Pipelines and Power: Towards a Political Ecology Discourse of Water in Southern Africa," in *Political Ecology: Science, Myth and Power*, ed. P. Stott and S. Sullivan, 132–153 (London: Edward Arnold, 2000).

376. R. Adler, N. Funke, K. Findlater, and A.R. Turton, *The Changing Relationship between the Government and the Mining Industry in South Africa: A Critical Assessment of the Far West Rand Dolomitic Water Association and the State Coordinating Technical Committee* (Pretoria: Council for Scientific and Industrial Research, 2007).

377. Turton, "South African Water."

378. P. J. Ashton, D. Hardwick, and C. M. Breen, "Changes in Water Availability and Demand within South Africa's Shared River Basins as Determinants of Regional Social-Ecological Resilience," in *Advancing Sustainability Science in South Africa*, ed. M. J. Burns and A. v. B. Weaver, 279–310 (Stellenbosch, South Africa: Stellenbosch University Press, 2008); P. Coetzer, "Water Management Under Pressure: South Africa Might be Facing a Water Contamination Crisis," *Achiever*, March 1, 2008; Turton, *Three Strategic Water Quality Challenges*.

379. See National Water Resource Strategy (NWRS), South African Department of Water Affairs and Forestry (DWAF), 2004 and P. J. Ashton, D. Hardwick, and C. M. Breen, "Changes in Water Availability and Demand Within South Africa's Shared River Basins as Determinants of Rgional Social-Ecological Resilience," in Advancing Sustainability Science in South Africa, eds. M. J. Burns and A.v.B. Weaver (Stellenbosch, South Africa: Stellenbosch University Press, 2008), 279–310.

380. Council for Scientific and Industrial Research, "Is the Glass Half Full or Half Empty?" *Science Scope* 3, no. 1 (2008): 18–19.

381. P. J. Oberholster and P. J. Ashton, *State of the Nation Report: An Overview of the Current Status of Water Quality and Eutrophication in South African Rivers and Reservoirs* (Pretoria: Council for Scientific and Industrial Research, 2008); Coetzer, "Water Management"; W. R. Harding and B. R. Paxton, *Cyanobacteria in South Africa: A Review*, Water Research

Commission Report No. TT 153/01 (Pretoria: Water Research Commission, 2001), 165; Turton, *Three Strategic Water Quality Challenges*.

382. South African Institute of Civil Engineers (SAICE), (presentation to Parliamentary Portfolio Committee on Water Affairs and Forestry, Cape Town, South Africa, June 4, 2008); Turton, *Three Strategic Water Quality Challenges*.

383. R. Adler, M. Claassen, L. Godfrey, and A. R. Turton, "Water, Mining and Waste: A Historical and Economic Perspective on Conflict Management in South Africa," *The Economics of Peace and Security Journal* 2, no. 2 (2007): 32–41; CSIR, *High Confidence Study of Children Potentially Affected by Radionuclide and Heavy Metal Contamination Arising from the Legacy of Mine Water Management Practices on the Far West Rand of South Africa*, Project Concept, Pretoria, South Africa; UN Office for the Coordination of Humanitarian Affairs, "South Africa: Paying the Price for Mining," IRIN Humanitarian News and Analysis, http://www.irinnews.org/Report.aspx?ReportId=76780; H. Coetzee, "Radioactivity and the Leakage of Radioactive Waste Associated with Witwatersrand Gold and Uranium Mining," in *Proceedings Uranium Mining and Hydrogeolog*, ed. B. J. Merkel, S. Hurst, E. P. Löhnert, and W. Struckmeier, 1–583 (Freiberg, Germany: GeoCongress, 1995); H. Coetzee, P. Wade, G. Ntsume, and W. Jordaan, *Radioactivity Study on Sediments in a Dam in the Wonderfonteinspruit Catchment. DWAF Report* (Pretoria: Department of Water Affairs and Forestry, 2002); H. Coetzee, P. Wade, and F. Winde, "Reliance on Existing Wetlands for Pollution Control Around the Witwatersrand Gold/Uranium Mines in South Africa—Are They Sufficient?" in *Uranium in the Aquatic Environment*, ed. B. J. Merkel, B. Planer-Friedrich, and C. Wolkersdorfer, 59–65 (Berlin: Springer, 2002); H. Coetzee, J. Venter, and G. Ntsume, *Contamination of Wetlands by Witwatersrand Gold Mines—Processes and the Economic Potential of Gold in Wetlands*, Council for Geosciences Report No. 2005-0106 (Pretoria: Council for Geosciences, 2005); H. Coetzee, F. Winde, and P. W. Wade, *An Assessment of Sources, Pathways, Mechanisms and Risks of Current and Potential Future Pollution of Water and Sediments in Gold-Mining Areas of the Wonderfonteinspruit Catchment*, WRC Report No. 1214/1/06 (Pretoria: Water Research Commission, 2006); P. W. Wade, S. Woodbourne, W. M. Morris, P. Vos, and N. W. Jarvis, *Tier 1 Risk Assessment of Selected Radionuclides in Sediments of the Mooi River Catchment*, WRC Project No. K5/1095 (Pretoria: Water Research Commission, 2002).

384. V. W. Werdmüller, "The Central Rand," in *Witwatersrand Gold—100 Years*, ed. E. S. A. Antrobus, 7–47 (Pretoria: The Geological Society of South Africa, 1986).

385. CSIR, *High Confidence Study of Children*; P. J. Hobbs and J. E. Cobbing, *A Hydrogeological Assessment of Acid Mine Drainage Impacts in the West Rand Basin, Gauteng Province*, CSIR Report No. CSIR/NRE/WR/ER/2007/0097/C. (Pretoria: CSIR/THRIP, 2007).

386. Natalie H. Aneck-Hahn, Gloria W. Schulenburg, Maria S. Bornman, Paulina Farias, and Christiaan de Jager, "Impaired Semen Quality Associated with Environmental DDT Exposure in Young Men Living in a Malaria Area in the Limpopo Province, South Africa," *Journal of Andrology* 28, no. 3 (2007): 423–434; M. S. Bornman, R. Delport, P. Becker, S. S. Risenga, and C. P. de Jager, "Urogenital Birth Defects in Neonates from a High-Risk Malaria Area in Limpopo Province, South Africa," *Epidemiology* 16, no. 5 (2005): S126–27.

387. F. G. Bell, S. E. T. Bullock, T. F. J. Hälbich, and P. Lindsay, "Environmental Impacts Associated with an Abandoned Mine in the Witbank Coalfield, South Africa," *International Journal of Coal Geology* 45 (2001): 195–216; J. N. Blignaut and N. A. King, "The Externality Cost of Coal Combustion in South Africa" (paper presented at the first annual Conference of the Forum for Economics and the Environment, Cape Town, South Africa, 2002); P. Hobbs, S. H. H. Oelofse, and J. Rascher, "Management of Environmental Impacts from Coal Mining in the Upper Olifants River Catchment as a Function of Age and Scale," ed. M. J. Patrick, J. Rascher, and A. R. Turton, special issue, *Reflections on Water in South Africa, Special Edition of International Journal of Water Resource Development* 24, no. 3 (2008): 417–432; F. D. I. Hodgson and R. M. Krantz, *Groundwater Quality Deterioration in the Olifants River Catchment above the Loskop Dam with Specialized Investigations in the Witbank Dam Sub-catchment*, Report No. 291/1/98 (Pretoria: Water Research Commission, 1998).

388. CSIR, *High Confidence Study of Children*; Hobbs and Cobbing, *A Hydrogeological Assessment*; K. Naicker, E. Cukrowska, and T. S. McCarthy, "Acid Mine Drainage from Gold Mining Activities in Johannesburg, South Africa, and Environs," *Environmental Pollution* 122 (2003): 29–40; F. Winde and I. J. Van Der Walt, "The Significance of Groundwater-Stream Interactions and Fluctuating Stream Chemistry on Waterborne Uranium Contamination of Streams—A Case Study from a Gold Mining Site in South Africa," *Journal of Hydrology* 287 (2004): 178–196; F. Winde, "Impacts of Gold-Mining Activities on Water Availability and

Quality in the Wonderfonteinspruit Catchment," in *An Assessment of Current and Future Water-Pollution Risk with Application to the Mooirivier-loop (Wonderfonteinspruit)*, ed. H. Coetzee, WRC Report No. K5/1214 (Pretoria: Water Research Commission, 2005), 14–38.

389. Harding and Paxton, *Cyanobacteria in South Africa*, 165; Oberholster and Ashton, *State of the Nation*; SAICE, "Presentation to Parliamentary Portfolio Committee."

390. Paton, "Dam Dirty," 32–39.

391. R. Spitz and M. Chaskalson, *The Politics of Transition: A Hidden History of South Africa's Negotiated Settlement* (Johannesburg: Witwatersrand University Press, 2000).

392. S. Johnston and A. Bernstein, *Voices of Anger: Protest and Conflict in Two Municipalities. Report to the Conflict and Governance Facility (CAGE)* (Johannesburg: The Centre for Development and Enterprise, 2007).

393. S. Nyathi, "No Water—No Vote," News 24, 2008, http://www.news24.com/News24/South_Africa/News/0,,2-7-1442_2326211,00.html.

394. R. Dixon, "Migrants Targeted for Fiery Deaths in South Africa," *Los Angeles Times*, May 20, 2008; N. Johnston and R. Wolmarans, "Xenophobic Violence Grips Johannesburg," *Mail and Guardian*, May 23, 2008, http://www.mg.co.za/articlePage.aspx?articleid=339509&area=/breaking_news/breaking_news__national/; B. Sibanda, "Falling Economic Standards Cause Xenophobia in South Africa," Afrik.com, May 14, 2008, http://en.afrik.com/article13586.html; P. Walker, "African Mobs Hunt Down Immigrants," *Guardian*, May 19, 2008.

395. Nyathi, "No Water."

396. A. Feinstein, *After the Party: A Personal and Political Journey inside the ANC* (Johannesburg: Jonathan Ball, 2007); Turton, "South African Water."

397. Turton, *Three Strategic Water Quality Challenges*.

398. Turton, "South African Water."

399. M. S. Basson, "South African Water Transfer Schemes and Their Impact on the Southern African Region," in *Water Resource Use in the Zambezi Basin. Proceedings of a Workshop held in Kasane, Botswana, 28 April–2 May 1993*, ed. T. Matiza, S. Craft, and P. Dale (Gland, Switzerland: IUCN, 1995); M. S. Basson, P. H. van Niekerk, and J. A. von Rooyen, *Overview of Water Resources Availability and Utilization in South Africa* (Pretoria: Department of Water Affairs and Forestry, 1997); Turton et al., "Gold, Scorched Earth and Water," 313–335.

400. CSIR, *High Confidence Study of Children*; Turton, "Three Strategic Water Quality Challenges, 2008.

401. Adam Weltz, "The Spy Who Came in from the Gold: Has a Leading Water Policy Expert Been Suspended for Being a Former Spook—or for Treading on the Mining Industry's Toes?" *Noseweek* 111 (January 2009): 30–31.
402. Ibid.
403. Ibid., 31.
404. Turton, "South African Water."
405. CSIR, *High Confidence Study of Children.*
406. M. S. Bornmanet al., "Urogenital Birth Defects," S126–27.
407. Aneck-Hahn et al., "Impaired Semen Quality"; Oberholster and Ashton, *State of the Nation Report.*
408. Oberholster and Ashton, *State of the Nation Report.*

CHAPTER 5

409. Nilanthi R. De Silva, Simon Brooker, Peter J. Hotez, Antonio Montresor, Dirk Engels, and Lorenzo Savioli, "Soil-Transmitted Helminth Infections: Updating the Global Picture," *Trends in Parasitology* 19, no. 12 (2003): 551.
410. UNAIDS, *Status of the Global HIV Epidemic, Report on the Global AIDS Epidemic*, 2008, 32, http://data.unaids.org/pub/GlobalReport/2008/jc1510_2008_global_report_pp29_62_en.pdf.
411. Alfred W. Crosby, *Ecological Imperialism: The Biological Expansion of Europe, 900–1900* (New York: Cambridge University Press, 1993), 63–67.
412. Andrew Davidson, *Hygiene and Diseases of Warm Climates* (Edinburgh: YJ Pentland, 1893).
413. Dennis Pirages, "Micro-Security: Disease Organisms and Human Well-Being," *ECSPR* 2 (1995): 9–14.
414. Thomas F. Homer-Dixon, "Environmental Scarcities and Violent Conflict: Evidence from Cases," *International Security* 19, no. 1 (1994): 5–40; Andrew Price-Smith, *Chaos and Contagion: Disease, Ecology, and National Security in the Era of Globalization* (Boston: MIT Press, 2009), 11–32.
415. National Intelligence Council, "National Intelligence Estimate: The Global Infectious Disease Threat and Its Implications for the United States," *ECSPR* 6 (2000): 33–65.
416. Valerie Brown, "Battle Scars: Global Conflicts and Environmental Health," *Environmental Health Perspectives* 112, no. 17 (2004): A994.
417. Alan Whiteside, Alex de Waal, and Tsadkan Gebre-Tensai, "AIDS, Security and the Military in Africa: A Sober Appraisal," *African Affairs* 105, no. 419 (2006): 201–218.

418. Rita Floyd, "Human Security and the Copenhagen School's Securitization Approach: Conceptualizing Human Security as a Securitizing Move," *Human Security Journal* 5 (winter 2007): 38–49.

419. RAND Corporation, *The Global Threat of New and Reemerging Infectious Diseases: Reconciling US National Security and Public Health Policy* (Santa Monica, CA: RAND, 2003), 1–12.

420. Harley Feldbaum, Preeti Patel, Egbert Sondorp, and Kelley Lee, "Global Health and National Security: The Need for Critical Engagement," *Medicine Conflict and Survival* 22 (2006): 192–198.

421. Benjamin Wisner, Piers M. Blaikie, Terry Cannon, and Ian Davis, *At Risk: Natural Hazards, People's Vulnerability and Disaster* (London: Routledge, 2005), 54–60.

422. Sue Lautze and Angela Raven-Roberts, "Violence and Complex Humanitarian Emergencies: Implications for Livelihoods Models," *Disasters* 30, no. 4 (2002): 383–401; Andre Le Sage and Nisar Majid, "The Livelihoods Gap: Responding to the Economic Dynamics of Vulnerability in Somalia," *Disasters* 26, no. 1 (2002): 10–27.

423. Gilberto Gallopín, "Linkages Between Vulnerability, Resilience, and Adaptive Capacity," *Global Environmental Change* 16 (2006): 293–303.

424. Nick Brooks, "Vulnerability, Risk and Adaptation: A Conceptual Framework," *Tyndall Centre for Climate Change Research* 38 (2003): 1–15.

425. Philip B. Adamson, "Dracontiasis in Antiquity," *Medical History* 32 (1988): 205.

426. Edmund Tapp, *Manchester Museum Mummy Project: Multidisciplinary Research on Ancient Egyptian Mummified Remains*, ed. Rosalie David (Manchester: Manchester University Press, 1979), 99.

427. Susan Watts, "An Ancient Scourge: The End of Dracunculiasis in Egypt," *Social Science and Medicine* 46, no. 7 (1998): 813–814.

428. Terence Walz, *Trade Between Egypt and Bilad as-Sudan: 1700–1820* (Cairo: Institut Français d'Archeologie Orientale, 1978).

429. WHO, "Dracunculiasis Eradication Initiative," 2009, 1, http://www.who.int/dracunculiasis/eradication/en/.

430. WHO, "Eradicating Guinea Worm Disease," 2008, 4, http://whqlibdoc.who.int/hq/2008/WHO_HTM_NTD_PCT_2008.1_eng.pdf.

431. Donald R. Hopkins and Craig P. Withers Jr., "Sudan's War and Eradication of Dracunculiasis," *The Lancet Supplement* 360 (2002): 22.

432. Ibid., 21.

433. Center for Disease Control, Division of Parasitic Diseases, "Fact Sheet for Dracunculiasis," 2008, http://www.cdc.gov/NCIDOD/DPD/PARA-SITES/dracunculiasis/factsht_dracunculiasis.htm.
434. Ibid.
435. Samuel Baron, ed., *Medical Microbiology*, 4th ed. (Galveston: University of Texas Press, 1996).
436. Ibid.
437. Doctors Without Borders, "Beyond Cholera: Zimbabwe's Worsening Crisis," 2009, http://www.doctorswithoutborders.org/publications/reports/2009/msf_beyond-cholera_zimbabwes-worsening-crisis.pdf; Physicians for Human Rights, "Health in Ruins: A Man-Made Disaster in Zimbabwe," 2009, http://physiciansforhumanrights.org/library/report-2009-01-13.html.
438. WHO, "Cholera in Zimbabwe: Update 2," 2009, http://www.who.int/csr/don/2009_02_20/en/index.html.
439. David Coltart, "A Decade of Suffering in Zimbabwe," CATO Institute report, 2008, http://physiciansforhumanrights.org/library/report-2009-01-13.html.
440. Ibid.; Physicians for Human Rights, "Health in Ruins."
441. John Zarocostas, "Aid Organisations Warn Zimbabwe's Cholera Crisis Is Far From Over," *British Medical Journal* 338 (2009): 693.
442. WHO, "Cholera in Zimbabwe: Update," December 26, 2008, http://www.who.int/csr/don/2008_12_26/en/index.html.
443. Clare Kapp, "Zimbabwe's Humanitarian Crisis Worsens," *The Lancet* 373 (2009): 447; WHO, "Cholera in Zimbabwe: Update 2."
444. Maire A. Connolly, Michelle Gayer, Michael J. Ryan, Peter Salama, Paul Spiegel, and David L. Heymann, "Communicable Diseases in Complex Emergencies: Impact and Challenges," *The Lancet* 364 (2004): 1974–1983.
445. Chad M. Briggs, Lucy Anderson, and Moneeza Walji, "Environmental Health Risks and Vulnerability in Post-conflict Regions," *Medicine, Conflict and Survival* 25, no. 2 (2009): 122–133.

CHAPTER 6

446. The views expressed are those of the author and not necessarily those of the Air Command and Staff College (ACSC).
447. Robert T. Watson, Marufu C. Zinyowera, and Richard H. Moss, eds., *The Regional Impact of Climate Change: An Assessment of Vulnerability* (Cambridge: Cambridge University Press, 1997).
448. M. Boko, I. Niang, A. Nyong, C. Vogel, A. Githeko, M. Medany, B. Osman-Elasha, R. Tabo, and P. Yanda, "2007: Africa," in *Climate Change*

2007: Impacts, Adaptation and Vulnerability. Contribution of Working Group II to the Fourth Assessment Report of the Intergovernmental Panel on Climate Change, ed. M. L. Parry, O. F. Canziani, J. P. Palutikof, P. J. van der Linden, and C. E. Hanson, 435 (Cambridge: Cambridge University Press, 2007).

449. Ibid.

450. Ibid., 448.

451. Rajendra K. Pachauri and Andy Reisinger, *Climate Change 2007: Synthesis Report. Contribution of Working Groups I, II and III to the Fourth Assessment, Report of the Intergovernmental Panel on Climate Change (IPCC)* (Geneva, Switzerland, 2007), 50.

452. Daniel C. Esty, M. A. Levy, C. H. Kim, A. de Sherbinin, T. Srebotnjak, and V. Mara, *2008 Environmental Performance Index* (New Haven, CT: Yale Center for Environmental Law and Policy, 2008), 8.

453. Ibid., 26.

454. Boko et al., "2007: Africa."

455. Ibid.

456. Hans-Martin Fussel and Richard J. T. Klein, "Climate Change Vulnerability Assessments: An Evolution of Conceptual Thinking," *Climatic Change* 75 (2006): 301–329.

457. Esty et al., *2008 Environmental Performance Index*, 19.

458. The Worldwatch Institute, ed., *State of the World 2009: Into a Warming World* (W. W. Norton: New York, 2009): 200.

459. M. L. Parry, O. F. Canziani, J. P. Palutikof, P. J. van der Inden, and C. E. Hanson, eds., *Climate Change 2007: Impacts, Adaptation and Vulnerability. Contribution of Working Group II to the Fourth Assessment Report of the Intergovernmental Panel on Climate Change* (Cambridge: Cambridge University Press, 2007), 871.

460. Fussel and Klein, "Climate Change Vulnerability Assessments," 313.

461. Parry et al., *Climate Change 2007: Impacts*, 875.

462. B. Metz, O. R. Davidson, P. R. Bosch, R. Dave, and L. A. Meyer, eds., *Climate Change 2007: Mitigation. Contribution of Working Group III to the Fourth Assessment Report of the Intergovernmental Panel on Climate Change* (Cambridge: Cambridge University Press, 2007), 814.

463. Fussel and Klein, "Climate Change Vulnerability Assessments," 314.

464. Parry et al., *Climate Change 2007: Impacts*, 878.

465. Ibid., 881.

466. Ibid., 883.

467. Ibid.

468. Esty et al., *2008 Environmental Performance Index.*

469. Ibid., 2.

470. Ibid., 41.

471. Ibid., 46.

472. Ibid., 19.

473. Ibid., 17–18.

474. Boko et al., *Climate Change 2007: Impacts*, 433–467.

475. United Nations Environment Programme (UNEP), *Global Environment Outlook: GEO 4 Environment for Development* (Valletta, Malta: Progress Press, 2007).

476. Nick Brooks, W. Neil Adger, and P. Mick Kelly, "The Determinants of Vulnerability and Adaptive Capacity at the National Level and the Implications for Adaptation," *Global Environmental Change* 15 (2005): 151.

477. Esty et al., *2008 Environmental Performance Index*, 20–23.

478. Ibid., 42.

479. Boko et al., *Climate Change 2007: Impacts*, 437–439.

480. Esty et al., *2008 Environmental Performance Index*, 44.

481. Boko et al., *Climate Change 2007: Impacts*, 437.

482. Esty et al., *2008 Environmental Performance Index*, 2008, 42–43.

483. Boko et al., *Climate Change 2007: Impacts*, 442.

484. Esty et al., *2008 Environmental Performance Index*, 47.

485. Boko et al., *Climate Change 2007: Impacts*, 444, 437.

486. Esty et al., *2008 Environmental Performance Index*, 49.

487. Boko et al., *Climate Change 2007: Impacts*, 435.

488. Esty et al., *2008 Environmental Performance Index*, 53.

489. Boko et al., *Climate Change 2007: Impacts*, 439, 442.

490. Esty et al., *2008 Environmental Performance Index*, 57.

491. E. S. Dwyer, S. Pinnock, J. M. Gregoire, and J. M. C. Pereira, "Global Spatial and Temporal Distribution of Vegetation Fire as Determined From Satellite Observations," *International Journal of Remote Sensing* 21 (2000): 1289–1302; P. I. Kempeneers, P. E. Swinnen, and F. Fierens, *GLOBSCAR Final Report*, TAP/N7904/FF/FR-001 version 1.2 (VITO, Belgium, 2000), 2002, http://geofront.vgt.vito.be/geosuccess/documents/Final%20Report.pdf;jsessionid=668127C51913A12DB357A8569F4FE74F.

492. Esty et al., *2008 Environmental Performance Index*, 61; Boris Worm, Edward B. Barbier, Nicola Beaumont, J. Emmett Duffy, Carl Folke, Benjamin S. Halpern, Jeremy B. C. Jackson et al., "Impacts of Biodiversity Loss on Ocean Ecosystem Services," *Science* 314, (2006): 787.

493. Boko et al., *Climate Change 2007: Impacts*, 435, 439.

494. Ibid., 448.
495. Esty et al., *2008 Environmental Performance Index*, 69–70.
496. Boko et al., *Climate Change 2007: Impacts*, 435.
497. Ibid., 441.
498. Esty et al., *2008 Environmental Performance Index*, 69; N. Stern, *The Economics of Climate Change: The Stern Review* (Cambridge: Cambridge University Press, 2007).
499. Boko et al., *Climate Change 2007: Impacts*, 441.
500. Esty et al., *2008 Environmental Performance Index*, 28.
501. Ibid., 26, 31.
502. Ibid., appendix C: Country Profiles, Egypt.
503. Ibid., 28.
504. Ibid., 31, 33.
505. UNEP, *Global Environmental Outlook*, 209.
506. Esty et al., *2008 Environmental Performance Index*, Appendix C: Country Profiles, Egypt.
507. Boko et al., *Climate Change 2007: Impacts*, 441–442.
508. Esty et al., *2008 Environmental Performance Index*, 28, 30.
509. Ibid., appendix C: Country Profiles, Nigeria.
510. Ibid., 31, 32.
511. Ibid., appendix C: Country Profiles, Nigeria.
512. Ibid.
513. Ibid., 28, 30.
514. Ibid., appendix C: Country Profiles, South Africa.
515. Ibid., 31, 32.
516. Ibid., appendix C: Country Profiles, South Africa.
517. Boko et al., *Climate Change 2007: Impacts*, 446.
518. UNEP, *Global Environmental Outlook*, 480.
519. G. Mutangadura, S. Ivens, and S. M. Donkor, "Assessing the Progress Made by Southern Africa in Implementing the MDG Target on Drinking Water and Sanitation," *Assessing Sustainable Development in Africa, Africa's Sustainable Development Bulletin*, Economic Commission for Africa, Addis Ababa (2005): 19–23.
520. UNEP, *Global Environmental Outlook*, 150.
521. Ibid., 177, 211.
522. Boko et al., *Climate Change 2007: Impacts*, 448.
523. Esty et al., *2008 Environmental Performance Index*, Appendix C: Country Profiles, South Africa.
524. Ibid., 28, 30.

525. Ibid., appendix C: Country Profiles, Kenya.

526. Ibid.

527. Ibid., 31–32.

528. Boko et al., *Climate Change 2007: Impacts*, 439.

529. Ibid., 442.

530. UNEP, *Global Environmental Outlook*, 93.

531. Esty et al., *2008 Environmental Performance Index*, appendix C: Country Profiles, Kenya.

532. Ibid., 30.

533. Ibid.

534. Boko et al., *Climate Change 2007: Impacts*, 445.

535. Esty et al., *2008 Environmental Performance Index*, appendix C: Country Profiles, Egypt.

536. Boko et al., *Climate Change 2007: Impacts*, 446.

CHAPTER 7

537. The views expressed are those of the author and not necessarily those of the U.S. Army War College or the Department of Defense.

538. Richard Matthew, "Environment and Security: Concepts and Definitions," in *Proceedings: Regional Asia Pacific Defence Environmental Workshop*, ed. Catherine Phinney and Kent Hughes Butts, 49 (Carlisle, PA: Center for Strategic Leadership, 1998).

539. U.S. Environmental Protection Agency, *Environmental Security: Strengthening National Security through Environmental Protection*, 160-F-99-001, September 1999, 1.

540. *National Security Strategy of the United States* (Washington, DC: Government Printing Office, 1988).

541. United Nations Environmental Programme, *From Conflict to Peace Building: The Role of Natural Resources and the Environment*, Nairobi, Kenya, February 2009, 8.

542. Ibid.

543. Ken Conca and Geoffrey D. Dabelko, eds., *The Case for Environmental Peacemaking* (Washington, DC: Woodrow Wilson Center Press, 2002).

544. Israel Ministry of Foreign Affairs, "The Multilateral Negotiations," *MFA Newsletter*, November 19, 2007.

545. Peter Gleick, ed., *Water in Crisis: A Guide to the World's Freshwater Resources* (New York: Oxford University Press, 1993), 93–94.

546. Larry Swatuk, "Environmental Cooperation for Regional Peace and Security in Southern Africa," in *Environmental Peacekeeping*, ed. K. Conca and G. Dabelko (Washington, DC: Woodrow Wilson Center Press, 2002), 120–160.

547. Bernard Griffard and Kent Hughes Butts, "South Asia Seismic Preparedness Conference," *CSL Issue Paper* (Carlisle, PA: U.S. Army War College, February 2005).

548. United Nations Development Programme, *Human Development Report 1994*, http:www.undp.org/en/reports/global.

549. U.S. Department of Defense, Directive 3000.05, *Military Support for Stability, Transition, and Reconstruction (SSTR) Operations*, November 2006.

550. Ambassador John Herbst, prepared statement before the Subcommittees on Oversight and Investigations and Terrorism and Unconventional Threats and Capabilities, House Committee on Armed Services, Washington, DC, February 26, 2008.

551. General George W. Casey, "The Strength of the Nation," *Army* (October 2007): 20–21.

552. U.S. Department of the Army, FM3-07, *Stability Operations*, October 2008.

553. National Commission on Terrorist Attacks upon the United States, *The 9/11 Commission Report* (New York: W.W. Norton, July 22, 2004), 378.

554. *National Security Strategy of the United States* (Washington, DC: Government Printing Office, 2002).

555. Hillary Clinton, Senate confirmation hearing transcript, January 13, 2009.

556. Datuk Seri Abdullah Achmad Badawi, speech at the United Malays National Organization Annual General Assembly, Kuala Lumpur, Malaysia, June 20, 2005.

557. William Green, "Family Comes Last," interview with Philippine President Gloria Macapagal Arroyo, *TIME Asia*, June 6, 2005, 27.

558. *National Strategy for Combating Terrorism* (Washington, DC: Government Printing Office, February 2003), 22.

559. Osama bin Laden, "Letter to the American People," *Observer*, November 24, 2002.

560. National Commission on Terrorist Attacks upon the United States, *The 9/11 Commission Report*, 376.

561. U.S. Department of State, *U.S. Government Counter Insurgency Guide*, January 2009, 7.

562. *National Strategy for Combating Terrorism*, 15–28.

563. Many Western concepts for development had little understanding of the environmental and social variables to African development. See, for exam-

ple, W.W. Rostow, *The Stages of Economic Growth: A Non-Communist Manifesto* (Cambridge: Cambridge University Press, 1960), 4–16.

564. Admiral Dennis C. Blair, *Annual Threat Assessment of the Intelligence Community for the Senate Select Committee on Intelligence, Statement for the Record*, February 2009, 34.

565. National Counterterrorism Center, *2009 Counterterrorism (CT) Calendar* (McLean, VA: National Counter Terrorism Center, 2009), 100.

566. Blair, *Annual Threat Assessment*, 6.

567. Angel Rabasa, *Radical Islam in East Africa* (Santa Monica, CA: RAND Corporation, 2009), 53–66. See also National Counterterrorism Center, *2009 Counterterrorism (CT) Calendar*, 104.

568. Anneli Botha, "Challenges in Understanding Terrorism in Africa," in *Understanding Terrorism in Africa: Building Bridges and Overcoming the Gaps*, ed. Anneli Botha and F. Wafula Okumu (Pretoria: Institute for Security Studies, 2008), 6–19.

569. Wallace C. Gregson, "Ideological Support: Attacking the Critical Linkage," *The Struggle Against Extremist Ideology*, ed. Kent Hughes Butts and Jeffrey C. Reynolds, 23 (Carlisle, PA: Center For Strategic Leadership, June 8, 2005).

570. Afrique en Ligne, "Ethiopian Food Aid Caught Up in Djibouti Port Congestion," February 27, 2009, www.afriqueenligne.com.

571. "Forum for Democratic Dialogue in Ethiopia—Forum," forum initiated by Arena, EDUM, OFDM, SDAF, UEDF (ESDP, OPC, and SEPDC), and Andenet, EthioSun.com, February 19, 2009.

572. Steve Bloomfield, "The Niger Delta: The Curse of the Black Gold," Global Policy Forum, New York, August 2, 2008. See also Amnesty International Canada, "Nigeria: Oil, Poverty and Violence," *Take Action*, August 9, 2006.

573. The Military Advisory Board, *National Security and the Threat of Climate Change*, Center for Naval Analysis Corporation, June 2007.

574. Michael Yao, "WHO Responding to the Drought Crisis in the Horn of Africa Region," (PowerPoint presentation, World Health Organization, May 17, 2006, http://www.who.int/hac/crises/international/hoafrica/sitreps/HoA_drought_crisis_17May2006.pdf).

575. Humanitarian Futures Programme for the Department for International Development, *Dimensions of Crises Impacts: Humanitarian Needs by 2015* (London: Kings College, January 17, 2007), 36–41.

576. bin Laden, "Letter." See also Rosalie Westenskow, "Intel Chief: Climate Change Threatens U.S. Security," United Press International, February 18, 2009, http://www.upi.com.

577. U.S. Environmental Protection Agency, *Environmental Security*, 1.

578. See chapter 8 for a further discussion of this point.

CHAPTER 8

579. The views expressed are those of the author and are not necessarily those of the African Center for Strategic Studies or the U.S. Department of Defense.

580. Callisto Madavo, *Africa: The Development Challenges of the 21st Century*, Woodrow Wilson International Center for Scholars, Africa Program Occasional Paper Series, September 2005.

581. Michael Ross, *Natural Resources and Civil War: An Overview with Some Policy Options*, (draft report prepared for conference on "The Governance of Natural Resources Revenues," World Bank and the Agence Francaise de Developpement, December 2002).

582. Ibid., 10.

583. Global Issues, *Conflicts in Africa: The Democratic Republic of Congo*, 2003, 4, http://www.globalissues.org/Geopolitics/Africa/DRC/2003.

584. Arvind Ganesan and Alex Vines, *Engine of War: Resources, Greed, and the Predatory State*, Human Rights Watch World Report, 2004, http://www.hrw.org/legacy/wr2k4/14.htm.

585. Global Issues, *Conflicts in Africa*, 4 (see note 576).

586. United Nations, *Report of the Panel of Experts on the Illegal Exploitation of Natural Resources and Other Forms of Wealth of the Democratic Republic of the Congo*, April 12, 2001, http://www.un.org/news/dh/latest/drcongo.html.

587. Dena Montague, "Stolen Goods: Coltan and Conflict in the Democratic Republic of Congo," *SAIS Review* 22, no. 1 (2002), 103–118.

588. BBC News, "Armies Strike at Uganda Rebels," December 14, 2008, http://news.bbc.co.uk/2/hi/africa/7782649.stm.

589. Global Witness, *The Sinews of War: Eliminating the Trade in Conflict Resources*, briefing document, November 2006, 5, http://www.globalpolicy.org/security/natres/generaldebate/2006/sinews.pdf.

590. Ibid.

591. Ibid., 6.

592. Ibid.

593. Global Policy Forum, *The Role of Liberia's Logging Industry on National and Regional Insecurity*, briefing to the UN Security Council, January 2001, 7.

594. John Markakis, *Environmental Degradation and Social Conflict in the Horn of Africa*, Environment and Conflicts Projects: ENCOP Occasional Papers, Center for Security Studies, Swiss Peace Foundation, 1995, 113.

595. International Crisis Group, *Report on Sudan*, http://www.crisisgroup.org/home/.

596. United Nations Environment Programme, "Environmental Degradation Triggering Tensions and Conflict in Sudan," news release, June 22, 2007.

597. Alex De Waal, "Counter-Insurgency on the Cheap," *London Review of Books* 26, no. 15, August 5, 2004, 3, http://www.lrb.co.uk/v26/n15/ waal11_.html.

598. Shane Bauer, "The Ecology of Genocide: The Darfur Crisis Has Environmental Roots," *The Environmental Magazine*, March–April 2007.

599. United Nations Environmental Programme, *Sudan: Post Conflict Environmental Assessment*, June 2007, http://www.unep.org/sudan.

600. Ochieng Kamudhayi, *Beyond the Peace Agreement: Threats and Challenges to Reconstruction and Peace in Somalia* (Kenya: Institute Diplomacy and International Studies, University of Nairobi), 31.

601. World Bank, *Conflict in Somalia: Drivers and Dynamics*, World Bank Economic and Sector Work Project Report, January 2005, 32.

602. Ibid.

603. Ibid.

604. Patricia Kameri Mbote, "From Conflict to Cooperation in the Management of Transboundary Waters: The Nile Experience," Linking Environment and Security papers, the Heinrich Böll Foundation North America, 12.

605. Ibid.

606. Patricia Kameri Mbote, *Environmental Conflict and Cooperation in The African Great Lakes Region: A Case Study of the Virungas*, Woodrow Wilson International Center for Scholars, December 2007, 42.

607. Ibid.

CHAPTER 9

608. The views expressed are those of the author and are not necessarily those of the U.S. Army War College or the Department of Defense.

609. United Nations Development Programme, *Human Development Report 1994: New Dimensions of Human Security*, 1994, 24–25, http://hdr.undp.org/en/reports/global/hdr1994/chapters/.

610. R. Kumchinsky Civil Society, Water Could Become Major Catalyst for Conflict, EURASIANET.org, http://icrc.org/departments/civilsociety/articles/pp091805.shtml.

611. International Committee of the Red Cross (ICRC), "War and Water," News Press Release 99/12, March 25, 1999, http://www.icrc.org/web/eng/siteeng0.nsf/htmlall/57jpmf?opendocument.
612. R. James Woolsey and Anne Korin, "How to Break Oil's Monopoly and OPEC's Cartel," *Innovations, Technology, Governance, Globalization* 3 (Fall 2008), http://www.mitpressjournals.org/toc/itgg/3/4.
613. United Nations Environmental Programme, "From Conflict to Peacebuilding."
614. Silja Halle, executive summary in United Nation Environmental Program's *From Conflict to Peacebuilding; The Role of Natural Resources and the Environment,* http://www.unep.org/pdf/pcdmb_policy_01.pdf.
615. Ibid.
616. United Nations General Assembly, *Report of the United Nations Conference on Environment and Development,* Rio de Janeiro, June 3–14, 1992, http://www.un.org/documents/ga/conf151/aconf15126-1annex1.htm.
617. United Nations Environment Programme, "Environment for Development," http://www.unep.org/Documents.Multilingual/Default.asp?DocumentID=43.
618. United Nations Environment Programme, "Environment for Development, Post Conflict and Development Branch," http://postconflict.unep.ch/index.php.
619. United Nations Environmental Program, "The Environment and Security Initiative; An International Partnership for Managing Conflict and Risk," 3, http://www.envsec.org/.
620. Ibid.
621. United Nations Peacekeeping Operations, *Principles and Guidelines,* 2008, 81–82, http://pbpu.unlb.org/PBPS/Library/Capstone_Doctrine_ENG.pdf.
622. United Nations Department of Peacekeeping Operations, *Environmental Policy for UN Field Missions,* draft document, March 2009; United Nations Department of Peacekeeping Operations, *Environmental Guidelines for UN Field Operations,* draft document, still in draft.
623. Lt. Col. (ret.) Brent C. Bankus, "Trip Report, ACOTA Training Mission of the 43rd South African Brigade, 31 January–14 February, 2009," 2.
624. NATO News, "NATO Naval Force Sets Sail for Africa," July 30, 2007, http://www.nato.int/docu/update/2007/07-july/e0730a.html; BBC News, "NATO Foils Somali Pirates' Attack," April 19, 2009, http://news.bbc.co.uk/2/hi/africa/8006827.stm.
625. Deniz Yuksel-Beten, *Summary Final Report of the Security Science Forum on Environmental Security,* NATO Headquarters, Brussels, Belgium, June

18, 2008, http:www.eapc_sps_d_2008_007.ssf report.pdf. Dr. Beten is the NATO Science for Peace and Security (SPS) Programme Section Head.

626. United Nations Security Council, "Security Council Authorizes Six-Month African Union Mission in Somalia, Unanimously Adopting Resolution 1744 (2007)" February 20, 2007, http://www.un.org/News/Press/docs/2007/sc8960.doc.htm.

627. Ramtane Lamara, "The United Nations Security Council on the Situation in Somalia," New York, December 16, 2008, http://www.africa-union.org/root/ua/Actualites/2008/dec/PSC/Statement%20by%20%20Lamamra%20on%20Somalia%2016-12-08.pdf.

628. United States Africa Command, "U.S. Africa Command," 1, http://www.africom.mil/AboutAFRICOM.asp.

629. United States Africa Command, "Partnering for Security and Stability" http://www.africom.mil/fetchBinary.asp?pdfID=20080930095216.

630. Author's experience in ACOTA South African training mission, CPX/CAX, January 30–February 14, 2009.

631. United States Africa Command, "Partnering for Security and Stability," http://www.africom.mil/fetchBinary.asp?pdfID=20080930095216.

632. Paul Collier, "Natural Resources and Conflict in Africa," *Crimes of War Project; War in Africa Crimes of War Project* (October 2004): 1–2, http://www.crimesofwar.org/africa-mag/afr_04_collier.html.

633. The United Nations, Department of Peacekeeping Organization, http://www.un.org/Depts/dpko/dpko/currentops.shtml#africa.

634. Food and Agriculture Organization of the United Nations, "Information on Fisheries Management in the Democratic Republic of the Congo," January 2001, http://www.fao.org/fi/oldsite/FCP/en/COD/BODY.HTM.

635. Interview with Lt. Col. Casey Englebrecht, S-3, operations officer for South African Defense Force Battalion in the DRC from 2006–2007, by the author at the 43rd Brigade HQs of the South African Defence Force in Pretoria, South Africa, February 10, 2009.

636. Halle, "From Conflict to Peacebuilding," 9.

637. Joint UNEP/OCHA Environment Unit, *Darfur Crisis, Rapid Environmental Assessment at the Kalma, Otash and Bajoum Camps* (Geneva, Switzerland: 2004), 4.

638. Ibid., 14.

CHAPTER 10

639. The views expressed are those of the author and not necessarily those of Air University, the Air Force, or the Department of Defense.

640. Kaplan, "The Coming Anarchy," 44–77; Thomas F. Homer-Dixon, "On the Threshold: Environmental Changes as Causes of Acute Conflict," *International Security* 16, no. 2 (1991): 76–116; Homer-Dixon, *Environmental Scarcity and Global Security*; Homer-Dixon, "Environmental Scarcities and Violent Conflict," 5–40; Homer-Dixon, *The Environment, Scarcity and Violence*; Thomas F. Homer-Dixon, "Debating Violent Environments," *Woodrow Wilson Institute Environmental Change and Security Project Report* 9 (2003): 89–96; Thomas F. Homer-Dixon and Jessica Blitt, eds., *Ecoviolence: Links Among Environment: Population and Security*, (Lanham, MD: Rowan and Littlefield, 1998); Ian Bannon and Paul Collier, eds., *Natural Resources and Violent Conflict: Options and Actions* (Washington, DC: The World Bank, 2003); Mats Berdal and David Malone, eds., *Greed and Grievance: Economic Agendas in Civil Wars* (Boulder, CO: Lynne Rienner Publishers, 2000).

641. Walter H. Kansteiner III and J. Stephen Morrison, *Rising U.S. Stakes in Africa: Seven Proposals to Strengthen U.S.-Africa Policy* (Washington, DC: Center for Strategic and International Studies, 2004), 88–103; Anthony Lake, Christine Todd Whitman, Princeton N. Lyman, and J. Stephen Morrison, *More than Humanitarianism: A Strategic U.S. Approach to Africa*, Independent Task Force Report No. 56, New York, Council on Foreign Relations, 2006, 28–76.

642. Daniel C. Esty, Marc Levy, Tanja Srebotnjak, and Alexander de Sherbinen, *Environmental Sustainability Index: Benchmarking National Environmental Stewardship* (New Haven, CT: Yale Center for Environmental Law and Policy, 2005); South African Department of Environmental Affairs and Tourism, "Overview of Integrated Environmental Management," Information Series O, 2004; *Botswana Vision 2016: Towards Prosperity For All* (Gaborone: Government Printing Office, 1997), 6–7.

643. *The Southern African Development Community Protocol on Shared Watercourse Systems*, (Gaborone: Southern African Development Community, 1995); Larry A. Swatuk, "Power and Water: The Coming Order in Southern Africa," in *The New Regionalism and the Future of Security and Development*, ed. B. Hettne, A. Inotai, and O. Sunkel, 210–247 (London: Palgrave, 2001); Anthony Turton, Peter Ashton, and Eugene Cloete, eds., *Transboundary Rivers, Sovereignty and Development: Hydropolitical Drivers in the Okavango River Basin* (Pretoria: University of Pretoria Centre for International Political Studies, 2003).

644. See http://www.un.org/genifo/bp/enviro.html for additional detail.

645. Alex Weaver, environmental scientist in the South African Council for Scientific and Industrial Research (CSIR), personal communications with the author, Pretoria, February 2005; Neil Carter, *The Politics of the Environment* (Cambridge: Cambridge University Press, 2001).

646. See, among many others, S. A. Marks, *The Imperial Lion: Human Dimensions of Wildlife Management in Africa* (Boulder, CO: Westview, 1984); Rodger Yeager and Norman N. Miller, *Wildlife, Wild Death: Land Use and Survival in Eastern Africa,* (Albany: State University of New York Press, 1986); David Anderson and Richard Grove, "The Scramble for Eden: Past, Present and Future in African Conservation," in *Conservation in Africa: People, Policies and Practice,* ed. D. Anderson and R. Grove, 1–12 (Cambridge: Cambridge University Press, 1987); Roderick P. Neumann, *Imposing Wilderness: Struggles over Livelihood and Nature Preservation in Africa* (Berkeley: University of California Press, 1998).

647. See, for example, Jonathan McShane and Thomas Adams, *The Myth of Wild Africa* (New York: Norton, 1992); J. Z. Z. Matowanyika, "Cast out of Eden: Peasants Versus Wildlife Policy in Savanna Africa," *Alternatives* 16, no. 1 (1989): 30–35; Valentine U. James, *Environmental and Economic Dilemmas of Developing Countries: Africa in the Twenty-first Century* (Westport, CT: Praeger, 1994); Urs P. Kreuter and Randy P. Simmons, "Who Owns the Elephant? The Political Economy of Saving the African Elephant," in *Wildlife in the Marketplace,* ed. T. Anderson and P. Hill, 147–165 (Lanham, MD: Rowan and Littlefield, 1995); Neumann, *Imposing Wilderness*; Clark Gibson, *Politicians and Poachers: The Political Economy of Wildlife Policy in Africa* (New York: Cambridge University Press, 1999); Dan Brockington, *Fortress Conservation: The Preservation of the Mkomazi Game Reserve, Tanzania* (Bloomington: Indiana University Press, 2002).

648. As examples, see William Ascher and Robert Healy, *Natural Resource Policymaking in Developing Countries* (Durham, NC: Duke University Press, 1990); Richard Grove, "Colonial Conservation, Ecological Hegemony and Popular Resistance: Toward a Global Synthesis," in *Imperialism and the Natural World,* ed. J. McKenzie, 15–50 (Manchester: Manchester University Press, 1990); Nancy Lee Peluso, *Rich Forests, Poor People: Resource Control and Resistance in Java* (Berkeley: University of California Press, 1993); Gary Gray, *Wildlife and People: The Human Dimensions of Wildlife Ecology* (Chicago: University of Illinois Press, 1993); Jan Dizard, *Going Wild: Hunting, Animal Rights and the Contested Meaning of Nature* (Amherst: University of Massachusetts Press, 1994); Mark

David Spence, *Dispossessing the Wilderness: Indian Removal and the Making of the National Parks* (Oxford: Oxford University Press, 1999); William M. Adams and Martin Mulligan, *Decolonizing Nature: Strategies for Conservation in a Post-Colonial Era* (London: Earthscan, 2003).

649. Barry Dalal-Clayton, David Brent, and Oliver Dubois, *Rural Planning in Developing Countries: Supporting Natural Resource Management and Sustainable Livelihoods* (London: Earthscan, 2003); Jeffrey Sayer and Bruce Campbell, *The Science of Sustainable Development: Local Livelihoods and the Global Environment* (Cambridge: Cambridge University Press, 2004); Frank Ellis and H. Ade Freeman, *Rural Livelihoods and Poverty Reduction Policies* (New York: Routledge, 2005); Monique Borgerhoff Mulder and Peter Coppolino, *Conservation: Linking Ecology, Economics and Culture* (Princeton, NJ: Princeton University Press, 2005).

650. Interview with Oliver Chapeyama, director of Enviroplan, Gaborone, Botswana, June 10, 2005. Chapeyama was a Harare-based employee of the U.S. Agency for International Development (USAID) in the late 1980s when the CBNRM program was first established in Zimbabwe, and had direct personal involvement in the U.S.-backed initiative.

651. Yemi Katerere, Sam Moyo, and Ryan Hill, *A Critique of Transboundary Natural Resource Management in Southern Africa*; Brian T. B. Jones, "Summary Report: Lessons Learned and Best Practices for CBNRM Policy and Legislation in Botswana, Malawi, Mozambique, Namibia, Zambia and Zimbabwe," WWF SARPO Regional Project for Community-Based Natural Resource Management (CBNRM) Capacity Building in Southern Africa, March 30, 2004, available at http://www.tnrf.org.

652. Interview with Oliver Chapeyama, director of Enviroplan, Gaborone, Botswana, June 10, 2005. For additional details on the Zimbabwe's CBNRM program, see Simon Metcalf, "Campfire: Zimbabwe's Communal Area Management Program for Indigenous Resources" (paper prepared for the Liz Claiborne and Art Ortenburg Foundation Workshop on Community-Based Conservation, July 1993); Brian Child, "Building the Campfire Paradigm—Helping Villagers Protect African Wildlife," Property and Environment Research Center (PERC) report, March 2004, http://www.perc.org/publications/percreports/june2004/paradigm.php.

653. For additional details on Zambia's CBNRM program, see R. Michael Wright, "Alleviating Poverty and Conserving Wildlife in Africa: An 'Imperfect' Model from Zambia," *National Association for the Practice of Anthropology (NAPA) Bulletin* 15 (1995).

654. Interview with Theodore Pierce, environmental officer and first secretary, U.S. Embassy Gaborone, June 11, 2005. For more detail, see Dan Henk, "The Environment, the US Military and Southern Africa," *Parameters* 36, no. 2 (summer 2006): 98–117.

655. Interviews with Tina Dooley-Jones, director of Technical Programs, USAID-Namibia, Windhoek, June 20, 2005; Antonie Esterhuizen, IRDNC co-coordinator, Kunene region and former employee of the Namibian Ministry of Environment and Tourism, Windhoek, June 22, 2005; and Malan Lindeque, permanent secretary of the Namibian Ministry of Environment and Tourism, Windhoek, June 20, 2005.

656. *Botswana Vision 2016.*

657. P. D. Tyson, "Climatic Change in Southern Africa: Past and Present Conditions and Possible Future Scenarios," *Climatic Change* 18 (1991): 158–258; Munyaradze Chenje and Phyllis Johnson, eds., *State of the Environment in Southern Africa* (Harare and Maseru: SARDC/SADC/IUCN, 1994); Rejoice Tsheko, "Rainfall Reliability, Drought and Flood Vulnerability in Botswana," July 23, 2003, http://www.wrc.org.za.

658. Daniel Schwartz and Ashbindu Singh, *Environmental Conditions, Resources and Conflicts*, (New York: United Nations Environmental Program, 1999); K. Seakle and B. Godschalk, "Waging War over Water in Africa," *Forum* 4 (2000): 110–133; Klare, *Resource War*; Swatuk, "Environmental Cooperation in Southern Africa," 143–145.

659. For detail on OKACOM, see particularly Isidro Pinheiro, Gabaake Gabaake, and Piet Heyns, "Cooperation in the Okavango River Basin: The OKACOM Perspective," in *Transboundary Rivers, Sovereignty and Development: Hydropolitical Drivers in the Okavango River Basin*, ed. A. Turton, P. Ashton, and E. Cloete, 105–118 (Pretoria: African Water Issues Unit, University of Pretoria, 2003).

660. Interview with Theodore Pierce, environmental officer and first secretary, U.S. Embassy Gaborone, June 11, 2005, and 2006; for additional insight, see Swatuk, "Environmental Cooperation."

661. Seakle and Godschalk, "Waging War over Water in Africa"; J. Ashton, "Are Southern African Water Conflicts Inevitable or Preventable?" in *Water War: An Enduring Myth or Impending Reality?*, ed. H. Solomon and A. Turton, 65–102 (Durban, South Africa: ACCORD Publishers, 2003); Turton, Ashton, and Cloete, eds., *Transboundary Rivers*.

662. Peace Parks Foundation, http://www.peaceparks.org.

663. Rosaleen Duffy, "Peace Parks: The Paradox of Globalization," *Geopolitics* 6, no. 2 (Autumn 2001): 1–26; Rosaleen Duffy, "The Potential and Pit-

falls of Global Environmental Governance: The Politics of Transfrontier Conservation Areas in Southern Africa," *Political Geography* 25, no. 1 (January 2006): 89–112; Manage Goodlae Uromi, Marc J. Stern, Cheryl Margolius, Ashley G. Lanfer, and Matthew Fladeland, eds., *Transboundary Protected Areas: The Viability of Regional Conservations Strategies* (New York: Food Products Press, 2003); Russell A. Mittermeier, Cyril F. Kormos, Christina Goettsch Mittermeier, Patricio Robles Gil, Trevor Sandwith, and Charles Besancon, *Transboundary Conservation: A New Vision for Protected Areas* (Washington, DC: Conservation International, 2005).

664. John Hanks, "Transfrontier Conservation Areas (TFCAs) in Southern Africa: Their Role in Conserving Biodiversity, Socioeconomic Development and Promoting a Culture of Peace," *Journal of Sustainable Forestry* 17, no. 1–2 (2003): 121–142; Raul P. Lejano, "Theorizing Peace Parks: Two Models of Collective Action," *Journal of Peace Research* 43, no. 5 (2006): 563–581.

665. Malcolm Draper, Marja Spierenburg, and Harry Wels, "African Dreams of Cohesion: Elite Pacting and Community Development in Transfrontier Conservation Areas in Southern Africa," *Culture & Organization* 10, no. 4 (December 2004): 341–353; Sanette Ferreira, "Problems Associated With Tourism Development in Southern Africa: The Case of Transfrontier Conservation Areas," *GeoJournal* 60, no. 3 (2004): 301–310; Marloes van Amerom and Bram Buscher, "Peace Parks in Southern Africa: Bringers of an African Renaissance?" *Journal of Modern African Studies* 43, no. 2 (2005): 159–182; Samantha Jones, "A Political Ecology of Wildlife Conservation in Africa," *Review of African Political Economy* 33, no. 109 (September 2006): 483–495.

666. Great Limpopo Transfrontier Park, http://www.greatlimpopopark.com; see also Anna Spenceley, "Tourism in the Great Limpopo Transfrontier Park," *Development Southern Africa* 23, no. 5 (December 2006): 649–667.

667. Peace Parks Foundation, http://www.peaceparks.org/news.

668. Interview with Werner Myburgh, KAZA project officer, Peace Parks Foundation, Stellenbosch, South Africa, April 18, 2006.

669. Interview with Theodore Pierce, environmental officer and first secretary, U.S. Embassy Gaborone, June 11, 2005, and 2006; interview with Malan Lindeque, permanent secretary of the Namibian Ministry of Environment and Tourism, Windhoek, June 20, 2005; interview and communications with Sedia Modise, Peace Parks Foundation representative in Botswana, between 2004 and 2006; interview with Lovemore Sola, Botswana representative of Conservation International, Maun, June 9, 2005.

670. Interview with Tina Dooley-Jones, director of Technical Programs, USAID-Namibia, Windhoek, June 20, 2005; interview with Malan Lindeque, permanent secretary of the Namibian Ministry of Environment and Tourism, Windhoek, June 20, 2005; interview and communications with Sedia Modise, Peace Parks Foundation representative in Botswana, between 2004 and 2006; interview with Werner Myburgh, KAZA project officer, Peace Parks Foundation, Stellenbosch, South Africa, April 18, 2006; interview with Theodore Pierce, environmental officer and first secretary, U.S. Embassy Gaborone, June 11, 2005, and 2006; interview with Lovemore Sola, Botswana representative of Conservation International, Maun, June 9, 2005; interview with Keith Kline, Regional National Resources Program manager, USAID Regional Center for Southern Africa, Gaborone, June 8, 2005; interview with Masego Madzwamuse, Botswana Country program coordinator, IUCN, Gaborone, June 10, 2005; interview with Hagen Moroney, commercial and economic officer, U.S. Embassy Gaborone, June 8, 2005; interview with Cornelis H. M. Vander Post, senior research fellow, Harry Oppenheimer Okavango Research Center, Maun, June 9, 2005.

671. Interview with Werner Myburgh, KAZA project officer, Peace Parks Foundation, Stellenbosch, South Africa, April 18, 2006; interview and communications with Sedia Modise, Peace Parks Foundation representative in Botswana, between 2004 and 2006; http://www.peaceparks.org/news, March 18, 2008.

672. Interview with Masego Madzwamuse, Botswana Country program coordinator, IUCN, Gaborone, June 10, 2005; interview with Lovemore Sola, Botswana representative of Conservation International, Maun, June 9, 2005.

673. USAID: Sub-saharan Africa, "Congo Basin Forest Partnership," http://www.usaid.gov/locations/sub-saharan_africa/initiatives/cbfp.html.

674. Interview with Dr. Lucas Gakale, permanent secretary of the Botswana Ministry of Environment, Wildlife and Tourism, Gaborone, March 7, 2004. For details, see Peter Gastrow, *Organised Crime in the SADC Region: Police Perceptions*, ISS Monograph 60 (Pretoria: Institute for Security Studies, 2001); Peter Gastrow, *Penetrating State and Business: Organised Crime in Southern Africa*, ISS Monographs 86 and 89, vols. 1 and 2 (Pretoria: Institute for Security Studies, 2003); Ted Leggett, ed., *Drugs and Crime in South Africa: A Study in Three Cities, Pretoria: Institute for Security Studies*, ISS Monograph 69 (Pretoria: Institute for Security Studies, 2002); *South African Crime Quarterly* (Pretoria: Institute for Security Studies), http://www.iss.co.za.

675. John Solomon, "The Danger of Terrorist Holes in Southern Africa," *Terrorism Monitor* 5, no. 15 (March 15, 2007), http://www.jamestown.org/terrorism/news/article.php2.issue_id=4038.

676. Obogonye Ochoche, "The Military and National Security in Africa," in *The Military and Militarism in Africa*, ed. E. Hutchful and A. Bathily (Dakar: CODESRIA Book Series, 1998); Herbert M. Howe, *Ambiguous Order: Military Forces in African States* (Boulder, CO: Lynne Rienner, 2001).

677. For a brief overview of the debate, see Jessica T. Mathews, "Redefining Security," *Foreign Affairs* 68, no. 2 (1989): 162–177; David Baldwin, "The Concept of Security," *Review of International Studies* 23, no. 1 (1997): 5–26; Barry Buzan, Ole Waever, and Jaap de Wilde, *Security: A New Framework for Analysis* (London: Lynne Rienner, 1998); Steve Smith, "The Increasing Insecurity of Security Studies: Conceptualizing Security in the Last Twenty Years," *Contemporary Security Policy* 20, no. 3 (1999): 72–101; Agostinho Zacarias, "Redefining Security," in *From Cape to Congo: Southern Africa's Evolving Security Challenges*, ed. M. Baregu and C. Lansberg, 31–52 (Boulder, CO: Lynne Rienner, 2003); Jakkie Cilliers, *Human Security in Africa: A Conceptual Framework for Review*, monograph for the African Human Security Initiative (Pretoria: Institute for Security Studies, 2004), available at: http://www.african review.org.

678. United Nations Development Programme, *Human Development Report, 1993*, http://www.undp.org/hdro/93.htm, and *Human Development Report, 1994*, http://www.undp.org/hdro/94.htm.

679. Daniel Deudney, "The Case Against Linking Environmental Degradation and National Security," *Millennium* 19, no. 3 (1990): 461–476; Daniel Deudney, "Environment and Security: Muddled Thinking," *Bulletin of the Atomic Scientists* 47, no. 3 (April 1991): 23–28; Jyrki Kakonen, ed., *Green Security or Militarized Environment?* (Aldershot, UK: Dartmouth, 1994); Marc A. Levy, "Is the Environment a National Security Issue?" *International Security* 20, no. 2 (Fall 1995): 35–62; Simon Dalby, "Contesting an Essential Concept: Reading the Dilemmas in Contemporary Security Discourse," in *Critical Security Studies*, ed. K. Krause and M. Williams, 3–32 (London: University College Press, 1997).

680. John Barnett, The *Meaning of Environmental Security: Ecological Politics and Policy in the New Security Era* (London: Zed Books, 2001), 157. See also Simon Dalby, *Environmental Security* (Minneapolis: University of Minnesota Press, 2002).

681. Simon Dalby, "Security, Modernity, Ecology: The Dilemmas of Post–Cold War Security Discourse," *Alternatives* 17, no. 1 (1992): 95–134; Gregory D. Foster, "Environmental Security: The Search for Strategic Legitimacy," *Armed Forces & Society* 27, no. 3 (spring 2001): 373–395; Larry A. Swatuk, "Environmental Security," in *Advances in the Study of International Environmental Politics*, ed. M. Betsill, K. Hochstetler, and D. Stevis (New York: Palgrave, 2006).

682. H. Solomon and M. van Aardt, eds., *Caring Security in Africa, Pretoria: Institute for Security Studies*, ISS Monograph 20 (Pretoria: Institute for Security Studies, 1998); Larry A. Swatuk and Peter Vale, "Why Democracy Is Not Enough: Southern Africa and Human Security in the Twenty-first Century," *Alternatives* 24, no. 3 (1999): 361–389; Keith Muloongo, Roger Kibasomba, and Jemima Njeri Kariri, *The Many Faces of Human Security: Case Studies of Seven Countries in Southern Africa* (Pretoria: Institute for Security Studies, 2005); Cheryl Hendricks, ed., *From State Security to Human Security in Southern Africa: Policy Research and Capacity Building Challenges*, ISS Monograph 122 (Pretoria: Institute for Security Studies, 2006).

683. See, for example, P. Godfrey Okoth, "The New East African Community: A Defense and Security Organ as the Missing Link," in *Africa at the Crossroads: Between Regionalism and Globalization*, ed. J. Mbaku and S. Saxena, 158–159 (Westport, CT: Praeger, 2004). Two years before the UN endorsed "human security," the celebrated Kampala Document published by the African Leadership Forum (and supported by the Organization of African Unity and United Nations) offered a definition of security very close to the later UN conceptualization. *The Kampala Document: Towards a Conference on Security, Stability, Development and Cooperation* (New York: African Leadership Forum, 1992), 9.

By the early twenty-first century, African scholars in seven nongovernmental research organizations across the continent had come together to form the African Human Security Initiative (AHSI), with a specific agenda of measuring the performance of African governments in promoting human security, insisting on accountability to human security principles (http://www.africanreview.org). By that point, the new African Union had agreed on a common African defense and security policy with a particularly nuanced and robust articulation of its human security foundations. See Jakkie Cilliers, *Human Security in Africa*.

684. See, particularly, Rocky Williams, "Defence in a Democracy: The South African Defence Review and the Redefinition of the Parameters of the

National Defence Debate," in *Ourselves to Know: Civil-Military Relations and Defence Transformation in Southern Africa*, ed. R. Williams, G. Cawthra, and D. Abrahams, 205–223 (Pretoria: Institute for Security Studies, 2003).

685. *South African White Paper on Defence, 1996* (Pretoria: 1 Military Printing Regiment of Department of Defence, 1998), 1.

686. Dan Henk, "The Botswana Defence Force and the War Against Poachers in Southern Africa," *Small Wars and Insurgencies* 16, no. 2 (June 2005): 170–191; Dan Henk, "Biodiversity and the Military in Botswana," *Armed Forces & Society* 32, no. 1 (2006): 273–291; Dan Henk, *The Botswana Defense Force and the Struggle for an African Environment* (New York: Palgrave Macmillan, 2007).

687. Nick Abel and Piers Blaike, "Elephants, People, Parks and Development: The Case of the Luangwa Valley, Zambia," *Environmental Management* 10 (1986): 735–751; Esmond Bradley Martin, "The Yemeni Rhino Horn Trade," *Pachyderm* 9 (1987): 13–17; Esmond Bradley Martin, "Report on the Trade in Rhino Products in Eastern Asia and India," *Pachyderm* 11 (1989): 14–22; Lucy Vigne and Esmond Bradley Martin, "Taiwan: The Greatest Threat to Africa's Rhinos," *Pachyderm* 11 (1989): 23–25; David Western, "The Undetected Trade in Rhino Horn," *Pachyderm* 11 (1989): 26–28; E. Barbier, J. Burgess, T. Swanson and D. Pearce, *Elephants, Economics and Ivory* (London: Earthscan, 1990); D. Balfour and S. Balfour, *Rhino—The Story of the Rhinoceros and a Plea for its Conservation* (Cape Town: Struik Publishers, 1991); Ian Douglas-Hamilton and Oria Douglas-Hamilton, *Battle for the Elephants* (New York: Viking Press, 1992); N. Leader-Williams, *The World Trade in Rhino Horn: A Review*, (TRAFFIC International, Species in Danger Series, 1992); Raymond Bonner, *At the Hand of Man: Peril and Hope for Africa's Wildlife* (New York: Alfred A. Knopf, 1993); Edward Ricciuti, "The Elephant Wars," *Wildlife Conservation* 96, no. 2 (1993): 14–35; Stephen Ellis, "Of Elephants and Men: Politics and Nature Conservation in South Africa," *Journal of Southern African Studies* 20, no. 1 (1994): 53–69; M. Sas-Rolfes, *Rhinos: Conservation, Economics and Trade-Offs* (London: Institute of Economic Affairs, 1995); Kathleen E. du Bois, "The Illegal Trade in Endangered Species," *African Security Review* 6, no. 1 (1997), http://www.iss.co.za.

688. Henk, "The Botswana Defence Force and the War Against Poachers"; Henk, "Biodiversity and the Military"; Henk, *The Botswana Defense Force and the Struggle for an African Environment*.

689. Thomas Schultheis, environmental officer on the staff of the U.S. European Command, Stuttgart-Vaihingen, Germany, communication with the author, February 24, 2005; interview with Lieutenant Colonel Brian P. Smith, U.S. Air Force, chief of the Office of Defense Cooperation, Pretoria, June 11, 2005; Colonel Seakle Godschalk, director of the Environmental Office in the Headquarters, South African National Defence Force, Pretoria, communications with the author, 2004–2006; see also K. Seakle and B. Godschalk, "Protecting Our Environment for a Quarter Century," *SA Soldier* 9, no. 10 (2002): 24–27.

690. Eddie Koch, "Nature Has the Power to Heal Old Wounds: War, Peace and Changing Patterns of Conservation in Southern Africa," in *South Africa in Southern Africa: Reconfiguring the region,* ed. D. Simon, 54–71 (London: James Currey, 1998); Peter J. Ashton, "Water Security for Multi-national River Basin States: The Special Case of the Okavango River," in *SIWI Seminar: Water Security for Multi-national River Basin States—Opportunity for Development,* ed. M. Falkenmark and J. Lundqvist (Stockholm: Swedish International Water Institute, 2000); Swatuk, "Power and Water," 210–247; Swatuk, "Environmental Cooperation in Southern Africa"; Turton et al., *Transboundary Rivers.*

691. Dan Henk, "The Environment, the US Military and Southern Africa," 111–113.

692. In fiscal years 1991 and 1993, the U.S. Congress authorized special allocations of US$15 million to "encourage African military establishments to [engage] in anti-poaching activities, wildlife protection and other efforts in support of Africa's environment." *U.S. Department of State Dispatch* 3, no.1 (January 1, 1992): 16, http://dosfan.lib.uic.edu/ERC/briefing/dispatch/1992/html/. Nambia, Zambia, and Zimbabwe were among the participants in these programs. Most of the funding was spent on equipment and training for antipoaching. The bulk of the materiel was delivered in 1993 and 1994, with some training and parts replacement continuing until about 1998. However, when the biodiversity purchases were completed, the U.S. military interest ended as well, and a decade later, it was hard to find much remaining evidence of the military "biodiversity" relationships.

693. Interview with Ninette Sadusky, Office of the Secretary of Defense (ODUSD I&E), Arlington, VA, March 14, 2005; interview with Lieutenant Jay Connors, U.S. Army, political-military staff officer in the Headquarters, U.S. European Command, Stuttgart-Vaihingen, Germany, May 21,

2006; Thomas Schultheis, environmental officer on the staff of the U.S. European Command, Stuttgart-Vaihingen, Germany, communication with the author, February 24, 2005. See also Dan Henk, "The Environment, the US Military, and Southern Africa," 105.

694. Interview with Ambassador Peter Chaveas, director of the Africa Center for Strategic Studies (ACSS), Washington, DC, April 4, 2006. See also http://www.africacenter.org and *The Bulletin (Africa Center for Strategic Studies)* 5, no. 1 (January 2007), http://www.dsca.osd.mil/newsletter/othrs-2007/acss-Jan2007.pdf.

695. Lauren Ploch, *CRS Report to Congress. Africa Command: U.S. Strategic Interests and the Role of the U.S. Military in Africa* (Washington, DC: Congressional Research Service, February 5, 2008); Major Shannon Beebe, U.S. Army, army staff, Washington, DC, multiple communications with the author, February–March 2007; multiple interviews with Stephen Burgess, associate professor and director of African Studies, Air War College, Maxwell AFB, AL, 2006–2009.

696. Abel Esterhuyse, "The Iraqization of Africa? Looking at AFRICOM from a South African perspective," *Strategic Studies Quarterly* 2, no. 1 (Spring 2008): 111–130; Theo Neethling, "Establishing AFRICOM: Pressing Questions, Political Concerns and Future Prospects," *Scientia Militaria* 36, no. 1 (2008): 31–51.

697. Mark Malan, "AFRICOM: A Wolf in Sheep's Clothing?" testimony to the U.S. Senate Foreign Relations Committee/Subcommittee on African Affairs, August 1, 2007.

698. AUSA, *Sustaining the Mission, Preserving the Environment, Securing the Future*, AUSA Torchbearer National Security Report, Arlington, VA, February 2007, http://www.ausa.org. For information on AEPI, see http://www.aepi.army.mil.

699. Interview with Geoffrey D. Dabelko, director, Environmental Change and Security Program, Woodrow Wilson International Center for Scholars, Washington DC, April 4, 2006. See also Woodrow Wilson Center Web site, http://www.wilsoncenter.org.

700. Interview with Malan Lindeque, permanent secretary of the Namibian Ministry of Environment and Tourism, Windhoek, June 20, 2005; interview with Theodore Pierce, environmental officer and first secretary, U.S. Embassy Gaborone, June 11, 2005, and 2006; interview with Jill Derderian, environment, science and technology officer, U.S. Embassy, Pretoria, June 8, 2005; interview with Arabang Kanego, staff officer, National Conservation Strategy Office Botswana Ministry of Environment, Wild-

life and Tourism, Gaborone, June 6, 2005; interview with Stevie Monna, staff officer, National Conservation Strategy Office Botswana Ministry of Environment, Wildlife and Tourism, Gaborone, June 6, 2005; interview with Nchunga Mushanana, staff officer, National Conservation Strategy Office Botswana Ministry of Environment, Wildlife and Tourism, Gaborone, June 6, 2005.

701. Kristiana Powell, *Opportunities and Challenges for Delivering on the Responsibility to Protect the African Union's Peace and Security Program*, ISS Monograph 119 (Pretoria: Institute for Security Studies, May 2005).

702. Swatuk, "Power and Water"; Swatuk, "Environmental Cooperation in Southern Africa"; Naison Ngoma, *Prospects for a Security Community Southern Africa* (Pretoria: Institute for Security Studies, June 2005); Cheryl Hendricks, ed., *From State Security to Human Security*.

703. Ken Conca, "The Case for Environmental Peacemaking," in *Environmental Peacemaking*, ed. K. Conca and G. Dabelko, 1–22 (Washington, DC: Woodrow Wilson Center Press, 2002); Ken Conca and Geoffrey D. Dabelko, "The Problems and Possibilities of Environmental Peacemaking," in *Environmental Peacemaking*, eds. K. Conca and G. Dabelko, 220–233 (Washington, DC: Woodrow Wilson Center Press, 2002).

704. Interview with Felix Monggae, CEO of the Kalahari Conservation Society, Gaborone, March 8, 2004; interview with Anthony Turton, environmental scientist at CSIR-Environmentek, Pretoria, June 22, 2004; interview with Masego Madzwamuse, Botswana Country program coordinator, IUCN, Gaborone, June 10, 2005; interview with Antonie Esterhuizen, IRDNC co-coordinator, Kunene region and former employee of the Namibian Ministry of Environment and Tourism, Windhoek, June 22, 2005; interview with Cornelis H. M. Vander Post, senior research fellow, Harry Oppenheimer Okavango Research Center, Maun, June 9, 2005; interview with Lovemore Sola, Botswana representative of Conservation International, Maun, June 9, 2005.

705. Interview with Gavin Cawthra, director, Center for Defence and Security Management, Graduate School of Public and Development Management, University of Witwatersrand, Johannesburg, June 22, 2004.

Chapter 11

706. The views expressed are those of the authors and not necessarily those of the U.S. Army or the Department of Defense.

707. Dan Henk, "Human and Environmental Security in Southern Africa: The Kavango-Zambezi (KAZA) Transfrontier Conservation Area Project" (paper presented at the 49th annual ISA convention, March 26, 2008).
708. Ibid.
709. Stephen Burgess, "Environment and Human Security in Africa" (paper presented at the 49th annual ISA convention, March 26, 2008).
710. Cindy Jebb, Laurel Hummel, Luis Rios, and Madelfia Abb, "Human and Environmental Security in the Sahel: A Modest Strategy for Success" (paper presented at the 49th annual ISA convention, March 26, 2008). See chapter 1 for a revised version of the paper.

BIBLIOGRAPHY

Abel, Nick, and Piers Blaike. "Elephants, People, Parks and Development: The Case of the Luangwa Valley, Zambia." *Environmental Management* 10 (1986): 735–751.

Abraham, Kinfe. "No Treaty to Prevent Ethiopia from Utilizing the Nile Water." *Addis Tribune*, February 25, 2005, http://www.addistribune. com/Archives/2005/02/25-02-05/No.htm.

Adams, William M., and Martin Mulligan. *Decolonizing Nature: Strategies for Conservation in a Post-Colonial Era*. London: Earthscan, 2003.

Adamson, Philip B. "Dracontiasis in Antiquity." *Medical History* 32 (1988): 204–209.

Adler, R., M. Claassen, L. Godfrey, and A. R. Turton. "Water, Mining and Waste: A Historical and Economic Perspective on Conflict Management in South Africa." *The Economics of Peace and Security Journal* 2, no. 2 (2007): 32–41.

Adler, R., N. Funke, K. Findlater, and A. R. Turton. *The Changing Relationship between the Government and the Mining Industry in South Africa: A Critical Assessment of the Far West Rand Dolomitic Water Association and the State Coordinating Technical Committee*. Pretoria: South African Council for Scientific and Industrial Research, 2007.

Africa Center for Strategic Studies. *The Bulletin* 5, no. 1 (January 2007), http://www.dsca.osd.mil/newsletter/othrs-2007/acss-Jan2007.pdf.

———. http://www.africacenter.org.

African Development Bank. *Policy for Integrated Water Resources Management*. April 2000. http://www.afdb.org/pls/portal/docs/PAGE/ADB_PAGEGROUP/TOPICS/ADBPROJECTSECTORS/WATER

SUPPLYAND SANITATION/ INTEGRATED_WATER_POLICY_
APR2000.PDF.

African Human Security Initiative (AHSI). http://www.africanreview.org.

African Unification Front. "Historical Overview: War In Africa: A Legacy of the Cold War." African Front.com, February 2009, http://www.africanfront.com/defense/defense4.php.

Alan, Nicol. "The Dynamics of River Basin Cooperation: The Nile and Okavango Basins." In *Transboundary Rivers, Sovereignty and Development: Hydropolitical Drivers in the Okavango River Basin*, edited by Anthony Turton, Peter Ashton, and Eugene Cloete, 267–286. Pretoria, South Africa: University of Pretoria, 2003.

Amnesty International Canada. "Nigeria: Oil, Poverty and Violence." *Take Action*, August 9, 2006.

Anderson, David, and Richard Grove. "The Scramble for Eden: Past, Present and Future in African Conservation." In *Conservation in Africa: People, Policies and Practice*, edited by D. Anderson and R. Grove, 1–12. Cambridge: Cambridge University Press, 1987.

Aneck-Hahn, Natalie H., Gloria W. Schulenburg, Maria S. Bornman, Paulina Farias, and Christiaan de Jager. "Impaired Semen Quality Associated with Environmental DDT Exposure in Young Men Living in a Malaria Area in the Limpopo Province, South Africa." *Journal of Andrology* 28, no. 3 (2007): 423–434.

Arab Human Development Report, Creating Opportunities for Future Generations. New York: United Nations Development Programme, 2002.

Ascher, William, and Robert Healy. *Natural Resource Policymaking in Developing Countries*. Durham, NC: Duke University Press, 1990.

Ashton, J. "Are Southern African Water Conflicts Inevitable or Preventable?" In *Water War: An Enduring Myth or Impending Reality?* edited by H. Solomon and A. Turton, 65–102. Durban, South Africa: ACCORD Publishers, 2003.

Ashton, Peter J. "Water Security for Multi-national River Basin States: The Special Case of the Okavango River." In *SIWI Seminar: Water Security for Multi-national River Basin States—Opportunity for Development*, edited by M. Falkenmark and J. Lundqvist. Stockholm: Swedish International Water Institute, 2000.

Ashton, P. J., D. Hardwick, and C. M. Breen. "Changes in Water Availability and Demand Within South Africa's Shared River Basins as Determinants of Regional Social-Ecological Resilience." In *Advancing Sustainability Science in South Africa*, edited by M. J. Burns and A. v. B. Weaver, 279–310. Stellenbosch, South Africa: Stellenbosch University Press, 2008.

Association of the United States Army. *Sustaining the Mission, Preserving the Environment, Securing the Future.* Torchbearer National Security Report. Arlington: AUSA, 2007. http://www.ausa.org.

Ayoob, Mohammed. *The Third World Security Predicament.* Boulder, CO: Lynne Rienner, 1995.

Badawi, Datuk Seri Abdullah Achmad. Speech at the United Malays National Organization Annual General Assembly. Kuala Lumpur, Malaysia, June 20, 2005.

Baldwin, David. "The Concept of Security." *Review of International Studies* 23, no. 1 (1997): 5–26.

Balfour, D., and S. Balfour. *Rhino—The Story of the Rhinoceros and a Plea for Its Conservation.* Cape Town: Struik Publishers, 1991.

Bankus, Lt. Col. (ret.) Brent C. "Trip Report, ACOTA Training Mission of the 43rd South African Brigade." January 31–February 31, 2009.

Barbier, E., J. Burgess, T. Swanson, and D. Pearce. *Elephants, Economics and Ivory.* London: Earthscan, 1990.

Barker, David Read. Address to Lake Tana Symposium, Bahir Dar University, Bahir Dar, Ethiopia, September 24, 2004, 1–6. http://www.worldlakes.org/uploads/Lake_Tana_ Symposium24Sep04_drb.pdf.

Barnett, John. *The Meaning of Environmental Security: Ecological Politics and Policy in the New Security Era*. London: Zed Books, 2001.

Baron, Samuel, ed. *Medical Microbiology*. 4th ed. Galveston: University of Texas Press, 1996.

Basson, M. S. "South African Water Transfer Schemes and Their Impact on the Southern African Region." In *Water Resource Use in the Zambezi Basin. Proceedings of a Workshop held in Kasane, Botswana, 28 April–2 May 1993*, edited by T. Matiza, S. Craft, and P. Dale. Gland, Switzerland: IUCN, 1995.

Basson, M. S., P. H. van Niekerk, and J. A. von Rooyen. *Overview of Water Resources Availability and Utilization in South Africa*. Pretoria: Department of Water Affairs and Forestry, 1997.

Batterbury, Simon, and Andrew Warren. "The African Sahel 25 Years After the Great Drought: Assessing Progress and Moving Towards New Agendas and Approaches." Special issue of *Global Environmental Change* 11, no. 1 (April 2001): 1–8.

Bauer, Shane. "The Ecology of Genocide: The Darfur Crisis Has Environmental Roots." *Environmental Magazine*, March–April 2007.

BBC News. "Crowds Cheer Senegal Peace Deal." December 30, 2004. http://news.bbc.co.uk/go/pr/fr/-/2/hi/africa/4135797.stm.

———. "Ministers Quit Over Senegal Tragedy." October 2, 2002. http://news.bbc.co.uk/2/hi/africa/2290879.stm.

Bega, S. "Water Expert Suspended." http://www.environment.co.za/topic.asp?TOPIC_ID=2179.

Beitler, Ruth M., and Cindy R. Jebb. "Egypt As a Failing State: Implications for U.S. National Security." *INSS Occasional Paper* 51, U.S. Air Force Academy (Colorado Springs, CO: July 2003).

Bell, F. G., S. E. T. Bullock, T. F. J. Hälbich, and P. Lindsay. "Environmental Impacts Associated With an Abandoned Mine in the Witbank Coalfield, South Africa." *International Journal of Coal Geology* 45 (2001): 195–216.

Bernstein, Lenny, Peter Bosch, Osvaldo Canziani, Zhenlin Chen, Renate Christ, Ogunlade Davidson, William Hare, Saleemul Huq, David Karoly, Vladimir Kattsov, Zbigniew Kundzewicz, Jian Liu, Ulrike Lohmann, Martin Manning, Taroh Matsuno, Bettina Menne, Bert Metz, Monirul Mirza, Neville Nicholls, Leonard Nurse, Rajendra Pachauri, Jean Palutikof, Martin Parry, Dahe Qin, Nijavalli Ravindra-nath, Andy Reisinger, Jiawen Ren, Keywan Riahi, Cynthia Rosenz-weig, Matilde Rusticucci, Stephen Schneider, Youba Sokona, Susan Solomon, Peter Stott, Ronald Stouffer, Taishi Sugiyama, Rob Swart, Dennis Tirpak, Coleen Vogel, and Gary Yohe. *Climate Change 2007: Synthesis Report. Contribution of Working Groups to the Fourth Assessment Report of the Intergovernmental Panel on Climate Change.* Cambridge: Cambridge University Press, 2007.

Berthold, Mark. *Linking Environment and Security: Conflict Prevention and Peacemaking in East and Horn of Africa.* Washington, DC: The Heinrich Böll Foundation North America. http://www.boell.org/docs/ E&S_Publication_Document_FINAL.pdf

bin Laden, Osama. "Letter to the American People." *Observer*, November 24, 2002.

Blair, Dennis C. *Annual Threat Assessment of the Intelligence Community for the Senate Select Committee on Intelligence.* February 2009.

Blignaut, J. N., and N. A. King. "The Externality Cost of Coal Combustion in South Africa." Paper presented at the first annual Conference of the Forum for Economics and the Environment, Cape Town, 2002.

Bloomfield Steve, "The Niger Delta: The Curse of the Black Gold." Global Policy Forum, August 2, 2008.

BNET Business Network. "National Security Strategy of the United States." *U.S. Department of State Bulletin*, April 1988. http://find-articles.com/p/articles/mi_m1079/is_n2133_v88/ai_6761415/pg_ 9?tag=content;col1.

Bodley, John H. *Anthropology and Contemporary Human Problems.* Mountain View, CA: Mayfield, 1996.

Boko, M., I. Niang, A. Nyong, C. Vogel, A. Githeko, M. Medany, B. Osman-Elasha, R. Tabo, and P. Yanda. "2007: Africa." In *Climate Change 2007: Impacts, Adaptation and Vulnerability. Contribution of Working Group II to the Fourth Assessment Report of the Intergovernmental Panel on Climate Change*, edited by M. L. Parry, O. F. Canziani, J. P. Palutikof, P. J. van der Linden and C. E. Hanson, 433–467. Cambridge: Cambridge University Press, 2007.

Bonner, Raymond. *At the Hand of Man: Peril and Hope for Africa's Wildlife*. New York: Alfred A. Knopf, 1993.

Borgerhoff Mulder, Monique, and Peter Coppolino. *Conservation: Linking Ecology, Economics and Culture*. Princeton, NJ: Princeton University Press, 2005.

Bornman, M. S., R. Delport, P. Becker, S. S. Risenga, and C. P. de Jager. "Urogenital Birth Defects in Neonates From a High-Risk Malaria Area in Limpopo Province, South Africa." *Epidemiology* 16, no. 5 (2005): S126–27.

Botha, Anneli. "Challenges in Understanding Terrorism in Africa." In *Understanding Terrorism in Africa: Building Bridges and Overcoming the Gaps*, edited by Wafula Okumu and Anneli Botha, 6–19. Pretoria: Institute for Security Studies, 2008.

Botswana Vision 2016: Towards Prosperity For All. Gaborone: Government Printing Office, 1997.

Brauch, Hans Günter, ed. *Security and Environment in the Mediterranean: Conceptualising Security*. Berlin: Springer, 2003.

Briggs, Chad M., Lucy Anderson, and Moneeza Walji. "Environmental Health Risks and Vulnerability in Post-conflict Regions." *Medicine, Conflict and Survival* 25, no. 2 (2009): 122–133.

Brockington, Dan. *Fortress Conservation: The Preservation of the Mkomazi Game Reserve, Tanzania*. Bloomington: Indiana University Press, 2002.

Brooks, Nick. "Vulnerability, Risk and Adaptation: A Conceptual Framework." *Tyndall Centre for Climate Change Research* [Working Paper 38 2003].

Brooks, Nick, W. Neil Adger, and P. Mick Kelly. "The Determinants of Vulnerability and Adaptive Capacity at the National Level and the Implications for Adaptation." *Global Environmental Change* 15 (2005): 151–163.

Brown, Lester. R. *Outgrowing the Earth: The Food Security Challenge in an Age of Falling Water Tables and Rising Temperatures*. New York: W. W. Norton, 2004.

———. *Plan B 2.0: Rescuing a Planet Under Stress and a Civilization in Trouble*. New York: W. W. Norton, 2006.

———. "Redefining National Security." Worldwatch Paper 14, 1977.

Brown, Michael E., Owen R. Cote Jr., Sean M. Lynn-Jones, and Steven E. Miller, eds., *New Global Dangers: Changing Dimensions of International Security*. Cambridge, MA: MIT University Press, 2004.

Brown, Oli, Anne Hammill, and Robert McLeman. "Climate Change as the 'New' Security Threat: Implications for Africa." *International Affairs* 83, no. 6 (2007): 1141–1154.

Brown, Valerie. "Battle Scars: Global Conflicts and Environmental Health." *Environmental Health Perspectives* 112, no. 17 (2004): A994.

Burgess, J., T. Swanson, and D. Pearce. *Elephants, Economics and Ivory*. London: Earthscan, 1990.

———. "Environment and Human Security in the Horn of Africa." *Journal of Human Security* 4, no. 2 (2008): 37–61.

Burgess, Stephen F. "African Security in the 21st Century: The Challenges of Indigenization and Multilateralism." *African Studies Review* 41, no. 2 (September 1998): 37–61.

Burns, M. J., and A. Weaver, eds. *Advancing Sustainability Science in South Africa*. Stellenbosch: Stellenbosch University Press, 2008.

Busby, Joshua W. *Climate Change and National Security: An Agenda for Action*. CSR-32. New York: Council on Foreign Relations, November 2007.

Butts, Kent Hughes. *The Struggle Against Extremist Ideology; Addressing the Conditions That Foster Terrorism*. Carlisle Barracks, PA: Center for Strategic Leadership, United States Army War College, 2005.

Buzan, Barry. *People, States, and Fear*. 2nd ed. Chapel Hill: University of North Carolina, 1983.

Buzan, Barry, Ole Waever, and Jaap de Wilde. *Security: A New Framework for Analysis*. London: Lynne Rienner, 1998.

Campbell, Kurt M., Jay Gulledge, J. R. McNeill, John Podesta, Peter Ogden, Leon Fuerth, R. James Woolsey, Alexander T. J. Lennon, Julianne Smith, and Richard Weitz. *The Ages of Consequences: The Foreign Policy and National Security Implications of Global Climate Change*. Washington, DC: Center for Strategic and International Studies, 2007.

Capra, Fritjof. *The Hidden Connections: Integrating the Biological, Cognitive, and Social Dimensions of Life into a Science of Sustainability*. New York: Doubleday, 2002.

———. *The Web of Life: A New Scientific Understanding of Living Systems*. New York: Anchor Books Doubleday, 1996.

Carius, Alexander, and Kurt M. Lietzmann. *Environmental Change and Security: A European Perspective*. Berlin: Springer Press, 1998.

Carter, Neil. *The Politics of the Environment*. Cambridge: Cambridge University Press, 2001.

Casey, General George W. "The Strength of the Nation." *Army* (October 2007): 20–30.

Catley, Andy, ed. *Livestock and Pastoral Livelihoods in Ethiopia: Impact Assessments of Livelihoods-Based Drought Interventions in Moyale*

and Dire Woredas. Pastoralist Livelihood Initiative, USAID and Feinstein International Center, Friedman School of Nutrition Science and
Policy, Tufts University, Addis Ababa, Ethiopia, 2007.

Cavalcanti, H. B. "Food Security." In *Human and Environmental Security: An Agenda for Change*, edited by Felix Dodds and Tim Pippard,
152–165. London: Earthscan, 2005.

Center for Disease Control, Division of Parasitic Diseases. "Fact Sheet
for Dracunculiasis." 2008. http://www.cdc.gov/NCIDOD/DPD/PAR-
ASITES/dracunculiasis/factsht_dracunculiasis.htm.

Chalk, Peter. "Case Study: The East African Corridor." In *Ungoverned
Territories: Understanding and Reducing Terrorist Risks*, edited by
Angel Rabasa, Steven Boraz, Peter Chalk, Kim Cragin, Theodore W.
Karasik, Jennifer D. P. Moroney, Kevin A. O'Brien, and John E. Peter-
set, 147–172. Santa Monica, CA: RAND Project AIR FORCE, 2007.
http://www.rnd.org/pubs/mongraphs/MG561/.

Chau, Donovan C. *US Counterterrorism in Sub-Saharan Africa: Understanding Costs, Cultures and Conflicts.* Carlisle, PA: Strategic Studies
Institute, 2008.

Chenje, Munyaradze, and Phyllis Johnson, eds. *State of the Environment in Southern Africa.* Harare and Maseru: SARDC/SADC/IUCN,
1994.

Child, Brian. "Building the Campfire Paradigm—Helping Villagers
Protect African Wildlife." Property & Environment Research Center (PERC) Report. March 2004. http://www.perc.org/publications/
percreports/june2004/paradigm.php.

Cilliers, Jakkie. *Human Security in Africa: A Conceptual Framework for
Review.* Monograph for the African Human Security Initiative. Pretoria: Institute for Security Studies, 2004. http:www.africanreview.org.

"Climate Change and International Security." Paper from the High Representative and the European Commission to the European Council,
S113/08, March 14, 2008. http://www.consilium.europa.eu/ueDocs/
cms_Data/docs/pressData/en/reports/99387.pdf.

Clinton, Hillary. Senate confirmation hearing transcript, January 13, 2009.

Clottney, Peter. "Ogaden Group Accuses Addis Ababa of Atrocities." VOANews.com, February 25, 2009.

Coetzee, H. "Radioactivity and the Leakage of Radioactive Waste Associated with Witwatersrand Gold and Uranium Mining." In *Proceedings Uranium Mining and Hydrogeolog*, edited by B. J. Merkel, S. Hurst, E. P. Löhnert, and W. Struckmeier, 1–583. Freiberg, Germany: GeoCongress, 1995.

Coetzee, H., F. Winde, and P. W. Wade. *An Assessment of Sources, Pathways, Mechanisms and Risks of Current and Potential Future Pollution of Water and Sediments in Gold-Mining Areas of the Wonderfonteinspruit Catchment.* WRC Report No. 1214/1/06. Pretoria: Water Research Commission, 2006.

Coetzee, H., J. Venter, and G. Ntsume. *Contamination of Wetlands by Witwatersrand Gold Mines—Processes and the Economic Potential of Gold in Wetlands.* Council for Geosciences Report No. 2005-0106. Pretoria: Council for Geosciences, 2005.

Coetzee, H., P. Wade, and F. Winde. "Reliance on Existing Wetlands for Pollution Control Around the Witwatersrand Gold/Uranium Mines in South Africa—Are They Sufficient?" In *Uranium in the Aquatic Environment*, edited by B. J. Merkel, B. Planer-Friederich, and C. Wolkersdorfer, 59–65. Berlin: Springer, 2002.

Coetzee, H., P. Wade, G. Ntsume, and W. Jordaan. *Radioactivity Study on Sediments in a Dam in the Wonderfonteinspruit Catchment.* DWAF Report. Pretoria: Department of Water Affairs and Forestry, 2002.

Coetzer, P. "Water Management Under Pressure: South Africa Might Be Facing a Water Contamination Crisis." *Achiever*, March 1, 2008.

Collier, Paul. "Natural Resources and Conflict in Africa." *Crimes of War Project: War in Africa*, October 2004. http://www.crimesofwar.org/africa-mag/afr_04_collier_print.html.

Collins, Robert O. *The Nile.* New Haven, CT: Yale University Press, 2002.

Coltart, David. *A Decade of Suffering in Zimbabwe.* CATO Institute Report. 2008. http://physiciansforhumanrights.org/library/report-2009-01-13.html.

Combined Joint Task Force—Horn of Africa. http://www.hoa.africom.mil/.

———. "Mission and Philosophy." http://www.hoa.centcom.mil/resources/english/facts.asp.

Conca, Ken. "The Case for Environmental Peacemaking." In *Environmental Peacemaking*, edited by K. Conca and G. Dabelko, 1–22. Washington, DC: Woodrow Wilson Center Press, 2002.

Conca, Ken, and Geoffrey Dabelko. *Environmental Peacekeeping.* Washington, DC: Woodrow Wilson Center Press, 2002.

———. "The Problems and Possibilities of Environmental Peacemaking." In *Environmental Peacemaking*, edited by K. Conca and G. Dabelko, 220–233. Washington, DC: Woodrow Wilson Center Press, 2002.

———, eds. *The Case for Environmental Peacemaking.* Washington, DC: Woodrow Wilson Center Press, 2002.

Connolly, Maire A., Michelle Gayer, Michael J. Ryan, Peter Salama, Paul Spiegel, and David L. Heymann. "Communicable Diseases in Complex Emergencies: Impact and Challenges." *The Lancet* 364 (2004): 1974–1983.

Council for Scientific and Industrial Relations. *High Confidence Study of Children Potentially Affected by Radionuclide and Heavy Metal Contamination Arising from the Legacy of Mine Water Management Practices on the Far West Rand of South Africa.* Project Concept Note. Pretoria: Council for Scientific and Industrial Research, February 26, 2008.

———. "Is the Glass Half Full or Half Empty?" *Science Scope* 3, no. 1 (2008): 18–19.

Creitaru, Ioana. "Environmental Security Seen from the European Union. The Case of EU Climate Policy as a Preventive Security Policy." Central and Eastern European Online Library. 2008. http://www.ceeol.com/.

Crosby, Alfred W. *Ecological Imperialism: The Biological Expansion of Europe, 900–1900.* New York: Cambridge University Press, 1993.

Curtin, Philip, Steven Feierman, Leonard Thompson, and Jan Vansina. *African History: From Earliest Times to Independence.* 2nd ed. London: Longman Group, 1995.

Dagne, Ted. *CRS Report for Congress, Africa: US Foreign Assistance Issues.* Washington, DC: Library of Congress, July 28, 2006.

———. *CRS Report for Congress, Africa and the War on Terrorism.* Washington, DC: Library of Congress, January 17, 2002.

Dalal-Clayton, Barry, David Brent, and Oliver Dubois. *Rural Planning in Developing Countries: Supporting Natural Resource Management and Sustainable Livelihoods.* London: Earthscan, 2003.

Dalby, Simon, "Contesting an Essential Concept: Reading the Dilemmas in Contemporary Security Discourse." In *Critical Security Studies*, edited by K. Krause and M. Williams, 3–32. London: University College Press, 1997.

———. *Environmental Security.* Minneapolis: University of Minnesota Press, 2002.

———. "Security, Modernity, Ecology: The Dilemmas of Post–Cold War Security Discourse." *Alternatives* 17, no. 1 (1992): 95–134.

Davenport, T. R. H., and C. Saunders. *South Africa: A Modern History.* Houndsmills, UK, and New York: Macmillan Press and St. Martin's Press, 2000.

David, A. Rosalie. *Manchester Museum Mummy Project.* Manchester, UK: Manchester University Press, 1988.

Davidson, Andrew. *Hygiene and Diseases of Warm Climates.* Edinburgh: YJ Pentland, 1893.

de Bruijn, Mirjam, and Han van Dijk. *Climate Variability and Political Insecurity: The Guera in Central Chad*. Voorstel Guera 00.4doc (unpublished), August 2002. http://www.ascleiden.nl/PDF/guera.pdf.

de Jong, Ferdinand, and Geneviève Gasser. "Contested Casamance/Discordante Casamance." *Canadian Journal of African Studies/Revue Canadienne des Études Africaines* 39, no. 2 (2005): 213–229.

Department of Water Affairs and Forestry (DWAF), Pretoria. *National Water Resource Strategy*. http://www.dwaf.gov.za/Documents/Politicies/NWRS/Default.html.

de Silva, Nilanthi R., Simon Brooker, Peter J. Hotez, Antonio Montresor, Dirk Engels, and Lorenzo Savioli. "Soil-Transmitted Helminth Infections: Updating the Global Picture." *Trends in Parasitology* 19, no. 12 (2003): 547–551.

De Waal, Alex. "Counter-Insurgency on the Cheap." *London Review of Books* 26, no. 15 (August 5, 2004). http://www.lrb.co.uk/v26/n15/waal11_.html.

Deudney, Daniel. "The Case Against Linking Environmental Degradation and National Security." *Millennium* 19, no. 3 (1990): 461–476.

———. "Environment and Security: Muddled Thinking." *Bulletin of the Atomic Scientists* 47, no. 3 (April 1991): 23–28.

Diouf, Mamadou. "Between Ethnic Memories and Colonial History in Senegal: The MFDC and the Struggle for Independence in Casamance." In *Ethnicity and Democracy in Africa*, edited by Bruce Berman, Dickson Eyoh, and Will Kymlicka, 218–239. Oxford: Oxford University Press, 2004.

Dixon, R. "Migrants Targeted for Fiery Deaths in South Africa." *Los Angeles Times*, May 20, 2008.

———. "Secret Lives of Servitude in Niger; The Government Has Banned Slavery and Denies It Exists. Though Few Speak of It, the Practice is a Tradition Many Do Not Question." *Los Angeles Times*, September 3, 2005, A2–A3.

Dizard, Jan. *Going Wild: Hunting, Animal Rights and the Contested Meaning of Nature*. Amherst: University of Massachusetts Press, 1994.

Doctors Without Borders. "Beyond Cholera: Zimbabwe's Worsening Crisis." 2000. http://www.doctorswithoutborders.org/publications/reports/2009/msf_beyond-cholera_zimbabwes-worsening-crisis.pdf.

Dodds, F., and T. Pippard, eds. *Human and Environmental Security: An Agenda for Change*. London: Earthscan, 2005.

Douglas-Hamilton, Ian, and Oria Douglas-Hamilton. *Battle for the Elephants*. New York: Viking Press, 1992.

Doyle, Mark. "Senegal's Southern Rebellion Starts To End." BBC News, December 16, 2004. http://news.bbc.co.uk/go/pr/fr/-/2/hi/Africa/4102267.stm.

Draper, Malcolm, Marja Spierenburg, and Harry Wels. "African Dreams of Cohesion: Elite Pacting and Community Development in Transfrontier Conservation Areas in Southern Africa." *Culture & Organization* 10, no. 4 (December 2004): 341–353.

du Bois, Kathleen E. "The Illegal Trade in Endangered Species," *African Security Review* 6, no. 1 (1997), http://www.iss.co.za.

Duffy, Rosaleen. "Peace Parks: The Paradox of Globalization," *Geopolitics* 6, no. 2 (autumn 2001): 1–26.

———. "The Potential and Pitfalls of Global Environmental Governance: The Politics of Transfrontier Conservation Areas in Southern Africa." *Political Geography* 25, no. 1 (January 2006): 89–112.

Duignan, Peter, and L. H. Gann. *The United States and Africa: A History*. Cambridge: Cambridge University Press, 1984.

Durant, Robert F. *The Greening of the US Military, Environmental Policy, National Security, and Organizational Change*. Washington, DC: Georgetown University Press, 2007.

Durbak, Christine K., and Claudia M. Strauss. "Securing a Healthier World." In *Human and Environmental Security: An Agenda for Change*, edited by Felix Dodds and Tim Pippard, 128–138. London: Earthscan, 2005.

Dwyer, E. S., S. Pinnock, J. M. Gregoire, and J. M. C. Pereira. "Global Spatial and Temporal Distribution of Vegetation Fire as Determined from Satellite Observations." *International Journal of Remote Sensing* 21 (2000): 1289–1302.

Easterly, William. *The White Man's Burden: Why the West's Efforts to Aid the Rest Have Done So Much Ill and So Little Good.* New York: Penguin, 2006.

Eberstadt, Nicholas. "The Future of AIDS." *Foreign Affairs* 81, no. 6 (November/December 2002): 22–45.

Edgecombe, R. "The Mfecane or Difaqane." In *New History of South Africa in Word and Image* (Afrikaans), edited by T. Cameron and S. B. Spies. Cape Town: Human and Rousseau, 1986.

Elbe, Stefan. "HIV/AIDS and the Changing Landscape of War." *International Security* 27, no. 2 (fall 2002): 159–177. Reprinted in *New Global Dangers: Changing Dimensions of International Security*, edited by E. Brown, Owen R. Cote Jr., Sean M. Lynn-Hones, and Steven E. Miller, 371–389. Cambridge, MA: MIT Press, 2004.

Elhance, Arun P. *Hydropolitics in the Third World: Conflict and Cooperation in International River Basins.* Washington, DC: United States Institute of Peace Press, 1999.

Ellis, Frank, and H. Ade Freeman. *Rural Livelihoods and Poverty Reduction Policies.* New York: Routledge, 2005.

Ellis, Stephen. "Briefing: The Pan-Sahel Initiative." *African Affairs* 103, no. 412 (2004): 459–464.

———. "Of Elephants and Men: Politics and Nature Conservation in South Africa." *Journal of Southern African Studies* 20, no. 1 (1994): 53–69.

Esterhuyse, Abel. "The Iraqization of Africa? Looking at AFRICOM from a South African Perspective." *Strategic Studies Quarterly* 2, no. 1 (spring 2008): 111–130.

Esty, Daniel C., M. A. Levy, C. H. Kim, A. de Sherbinin, T. Srebotnjak, and V. Mara. *2008 Environmental Performance Index*. New Haven, CT: Yale Center for Environmental Law and Policy, 2008.

Esty, Daniel C., M. Levy, Tanja Srebotnjak, and Alexander de Sherbinen. *2005 Environmental Sustainability Index: Benchmarking National Environmental Stewardship*. New Haven, CT: Yale Center for Environmental Law and Policy, 2005.

"Ethiopian Food Aid Caught Up in Djibouti Port Congestion." *Afrique en Ligne, Djibouti*, February 27, 2009. http://www.afriquejet.com/news.html.

Evans, M. M. *The Boer War: South Africa 1899–1902. Osprey Military*. Mechanicsburg, PA: Stackpole Books, 1999.

Faris, Stephan. "The Real Roots of Darfur." *Atlantic Monthly*, April 2007.

Fawcett, M. H. *The Concentration Camps in South Africa*. London: Westminster Gazette, 1901.

Federal News Service. "Habbanae Loans of North Niger for Harmony List." June 20, 2006. http://0-proquest.umi.com.usmalibrary.usma.edu.

Feinstein, A. *After the Party: A Personal and Political Journey inside the ANC*. Johannesburg: Jonathan Ball, 2007.

Feinstein, Lee, and Anne-Marie Slaughter. "A Duty to Prevent." *Foreign Affairs* 83, no. 1 (January/February 2004): 136–150.

Feldbaum, Harley, Preeti Patel, Egbert Sondorp, and Kelley Lee. "Global Health and National Security: The Need for Critical Engagement." *Medicine Conflict and Survival* 22 (2006): 192–198.

Ferreira, Sanette. "Problems Associated with Tourism Development in Southern Africa: The Case of Transfrontier Conservation Areas." *GeoJournal* 60, no. 3 (2004): 301–310.

Floyd, Rita. "Human Security and the Copenhagen School's Securitization Approach: Conceptualizing Human Security as a Securitizing Move." *Human Security Journal* 5 (winter 2007): 38–49.

Food and Agriculture Organization of the United Nations. "Information on Fisheries Management in the Democratic Republic of the Congo." January 2001. http://www.fao.org/fi/oldsite/FCP/en/COD/BODY.HTM.

Forest, James. *The Making of a Terrorist: Recruitment, Rtaining, and Root Causes.* 3 Vols 2005–06. Praeger International Security. http://psi.praeger.com.

"Forum for Democratic Dialogue in Ethiopia—Forum." Forum initiated by Arena, EDUM, OFDM, SDAF, UEDF (ESDP, OPC, and SEPDC), and Andenet. EthioSun.com, February 19, 2009.

Foster, Gregory D. "Environmental Security: The Search for Strategic Legitimacy." *Armed Forces & Society* 27, no. 3 (spring 2001): 373–395.

Foucher, Vincent. "Senegal: The Resilient Weakness of Casamançais Separatists." In *African Guerrillas: Raging Against the Machine*, edited by Morten Bøås and Kevin C. Dunn, 171–197. Boulder, CO: Lynne Rienner, 2007.

Fukuyama, Francis. *State-Building: Governance and World Order in the 21st Century.* Ithaca, NY: Cornell University Press, 2004.

Fussel, Hans-Martin, and Richard J. T. Klein. "Climate Change Vulnerability Assessments: An Evolution of Conceptual Thinking." *Climatic Change* 75 (2006): 301–329.

Gallopín, Gilberto. "Linkages Between Vulnerability, Resilience, and Adaptive Capacity." *Global Environmental Change* 16 (2006): 293–303.

Ganesan, Arvind, and Alex Vines. *Engine of War: Resources, Greed, and the Predatory State.* Human Rights Watch Report. 2004. http://www.hrw.org/legacy/wr2k4/14.htm.

Gastrow, Peter. *Organised Crime in the SADC Region: Police Perceptions.* ISS Monograph 60. Pretoria: Institute for Security Studies, 2001.

————. *Penetrating State and Business: Organised Crime in Southern Africa.* ISS Monographs 86 and 89. Pretoria: Institute for Security Studies, 2003.

Gazibo, Mamoudou. "Foreign Aid and Democratization: Benin and Niger." *African Studies Review* 48, no. 3 (2005): 8–9. http://0-proquest. umi.com.usmalibrary.usma.edu.

Gibson, Clark. *Politicians and Poachers: The Political Economy of Wildlife Policy in Africa.* New York: Cambridge University Press, 1999.

Gleditsch, Nils Petter. "Armed Conflict and the Environment: A Critique of the Literature." *Journal of Peace Research* 35, no. 3 (1998): 381–400.

Gleick, Peter. "Environment and Security: The Clear Connections." *The Bulletin of Atomic Scientists* 47 (1991): 17–21.

————. "Water Conflict Chronology." Pacific Institute for Studies in Development, Environment, and Security. December 1, 2004. http://www.worldwater.org/conflict.htm.

————, ed. *Water in Crisis: A Guide to the World's Freshwater Resources.* New York: Oxford University Press, 1993.

Glenn, Jerome C., Theodore J. Gordon, Renet Sills Perelet, and Joe B. Sills. Millennium Project. *Environmental Security: United Nations Doctrine for Managing Environmental Issues in Military Actions.* U.S. Army Policy Institute. 1999.

Global Policy Forum. *The Role of Liberia's Logging Industry on National and Regional Insecurity.* Briefing to the UN Security Council. January 2001. http://www.globalpolicy.org/security/issues/liberia/report/gwtimber. htm.

Global Risk Outlook 2006, 2005. London: Exclusive Analysis Ltd, 2005–2006.

Global Water Partnership East Africa. *Partnership for Africa's Water Development 1 Program.* (PWPD), Q1 Progress Report submitted to the Global Water Partnership Organization, Stockholm, Sweden, 2005.

http://www.gwpena.org/dwn/PAWD%201_2005Annual%20report%20_GWP%20EnA.pdf.

Global Witness. *The Sinews of War: Eliminating the Trade in Conflict Resources*. November 2006. http://www.globalpolicy.org/security/natres/generaldebate/2006/sinews.pdf.

Godschalk, Seakle K. B. "Protecting Our Environment for a Quarter Century." *SA Soldier* 9, no. 10 (2002): 24–27.

Grada, Cormac Ó. "Making Famine History." *Journal of Economic Literature* 45 (March 2007): 3–36.

Gray, Gary. *Wildlife and People: The Human Dimensions of Wildlife Ecology*. Chicago: University of Illinois Press, 1993.

Gray, John Milner. *A History of the Gambia*. London: Frank Cass, 1966.

Great Limpopo Transfrontier Park. http://www.greatlimpopopark.com.

Green, William. "Family Comes Last." Interview with Philippine President Gloria Macapagal Arroyo. *TIME Asia*, June 6, 2005.

Gregson, Wallace C. "Ideological Support: Attacking the Critical Linkage." In *The Struggle Against Extremist Ideology*, edited by Kent Hughes Butts and Jeffrey C. Reynolds, 20–29. Carlisle, PA: Center for Strategic Leadership, June 8, 2005.

Griffard, Bernard, and Kent Hughes Butts. "South Asia Seismic Preparedness Conference." CSL Issue Paper. Carlisle, PA: US Army War College, February 2005.

Grove, Richard. "Colonial Conservation, Ecological Hegemony and Popular Resistance: Toward a Global Synthesis." In *Imperialism and the Natural World*, edited by J. McKenzie, 15–50. Manchester: Manchester University Press, 1990.

Gurr, Tedd. "Why Minorities Rebel: Explaining Ethnopolitical Protest and Rebellion." In *Minorities and Risk: A Global View of Ethnopolitical Conflicts*, 123–138. Washington, DC: United States Institute of Peace, October 1997.

Gutteridge, W. "Mineral Resources and National Security." In *Conflict Studies* 162 (1984): 1–25. Reprinted in *South Africa: From Apartheid to National Unity, 1981–1994*, edited by W. Gutteridge, 59–83. Aldershot, Hants, UK and Brookfield, VT: Dartmouth Publishing, 1995.

Halle, Silja. Executive summary. United Nation Environmental Program's *From Conflict to Peacebuilding; The Role of Natural Resources and the Environment.* Nairobi, Kenya, UNEP, 2009.

Halperin, Morton. H. "Guaranteeing Democracy." *Foreign Policy 91* (summer 1993): 105–122.

Hanks, John. "Transfrontier Conservation Areas (TFCAs) in Southern Africa: Their Role in Conserving Biodiversity, Socioeconomic Development and Promoting a Culture of Peace." *Journal of Sustainable Forestry* 17, no. 1–2 (2003): 121–142.

Harding, W. R., and B. R. Paxton. *Cyanobacteria in South Africa: A Review.* Water Research Commission Report No. TT 153/01. Pretoria: Water Research Commission, 2001.

Hasian, Marouf. "The 'Hysterical' Emily Hobhouse and Boer War Concentration Camp Controversy." In *Western Journal of Communication*, March 2003. http://www.accessmylibrary.com/coms2/summary_0286-23546431_ITM.

Heath, David E. "Niger: More Than a Food Shortage." *Environment* 47, no. 10 (December 2005): 1. http://0proquest.umi.com.usmalibrary.usma.edu.

Hendricks, Cheryl, ed. *From State Security to Human Security in Southern Africa: Policy Research and Capacity Building Challenges.* ISS Monograph 122 (April). Pretoria: Institute for Security Studies, 2006.

Henk, Dan. "Biodiversity and the Military in Botswana," *Armed Forces and Society* 32, no. 1 (2006): 273–291.

———. *The Botswana Defense Force and the Struggle for an African Environment.* New York: Palgrave Macmillan, 2007.

————. "The Botswana Defence Force and the War Against Poachers in Southern Africa." *Small Wars and Insurgencies* 16, no. 2 (June 2005): 170–191.

————. "The Environment, the US Military and Southern Africa." *Parameters* 36, no. 2 (summer 2006): 98–117.

Herbst, Ambassador John. Prepared statement before the Subcommittees on Oversight and Investigations and Terrorism and Unconventional Threats and Capabilities, House Committee on Armed Services, Washington DC, February 26, 2008.

Hill, Ginny. "Military Focuses on Development in Africa: In Djibouti, US Forces Combat Terrorism with Civil Affairs Work. Will This Be a Model for a Future US Military Command in Africa?" *Christian Science Monitor*, June 22, 2007.

Hinkley, John C. *Global Warming and Africa—Where Engagement Can Make a Difference.* Carlisle, PA: US Army War College, 2007.

Hobbs, P. J., and J. E. Cobbing. *A Hydrogeological Assessment of Acid Mine Drainage Impacts in the West Rand Basin, Gauteng Province.* CSIR Report No. CSIR/NRE/WR/ER/2007/0097/C. Pretoria: CSIR/THRIP, 2007.

Hobbs, P. J., S. H. H. Oelofse, and J. Rascher. "Management of Environmental Impacts from Coal Mining in the Upper Olifants River Catchment as a Function of Age and Scale." Edited by M. J. Patrick, J. Rascher, and A. R. Turton. *Reflections on Water in South Africa*, special issue *International Journal of Water Resource Development* 24, no. 3 (2008): 417–432.

Hobhouse, E. *1901 Report of a Visit to the Camps of Women and Children in the Cape and Orange River Colonies.* London: Friars Printing Association, 1907.

Hodgson, F. D. I., and R. M. Krantz. *Groundwater Quality Deterioration in the Olifants River Catchment above the Loskop Dam with*

Specialized Investigations in the Witbank Dam Sub-catchment. Report No. 291/1/98. Pretoria: Water Research Commission, 1998.

Hoffman, Bruce. "Terrorism." Lecture to Terrorism and Counterterrorism class, USMA, West Point, April 2004.

Homer-Dixon, Thomas F. "Debating Violent Environments." *Woodrow Wilson Institute Environmental Change and Security Project Report* 9 (2003): 89–96.

———. *Environment, Scarcity, and Violence.* Princeton, NJ: Princeton University Press, 1999.

———. *Environmental Scarcity and Global Security.* New York: Foreign Policy Association, 1993.

———. "Environmental Scarcities and Violent Conflict." In *New Global Dangers: Changing Dimensions of International Security*, edited by Michael E. Brown et al., 301–336. Cambridge, MA: MIT, 2004.

———. "Environmental Scarcities and Violent Conflict: Evidence From Cases." *International Security* 19, no. 1 (1994): 5–40.

———. "On the Threshold: Environmental Changes as Causes of Acute Conflict." *International Security* 16, no. 2 (1991): 76–116.

Homer-Dixon, Thomas F., Ian Bannon, and Paul Collier, eds. *Natural Resources and Violent Conflict: Options and Actions.* Washington, DC: World Bank, 2003.

Homer-Dixon, Thomas F., and Jessica Blitt, eds. *Ecoviolence: Links Among Environment, Population and Security.* Lanham, MD: Rowan and Littlefield, 1998.

Homer-Dixon, Thomas F., Mats Berdal, and David Malone, eds. *Greed and Grievance: Economic Agendas in Civil Wars.* Boulder, CO: Lynne Rienner, 2000.

Hopkins, Donald R., and Craig P. Withers Jr. "Sudan's War and Eradication of Dracunculiasis." *The Lancet Supplement* 360 (2002): 21–22.

Howe, Herbert M. *Ambiguous Order: Military Forces in African States.* Boulder, CO: Lynne Rienner, 2001.

Hulme, Mike, and P. M. Kelly. "Exploring the Links Between Desertification and Climate Change." *Environment* 35, no. 6 (1993): 4–11, 39–45.

Human Security Doctrine for Europe. Barcelona Report of the Study Group on Europe's Security Capabilities, September 15, 2004.

Humanitarian Futures Programme for the Department for International Development. *Dimensions of Crises Impacts: Humanitarian Needs by 2015.* London: Kings College, January 17, 2007.

Hummel, Laurel J. "Lowering Fertility Rates in Developing States: Security and Policy Implications for Sub-Saharan Africa." Master of Strategic Studies (M.S.S.) thesis, Carlisle Barracks, PA: United States Army War College, 2006.

Hunt, Emily. "Al-Qaeda's North African Franchise: The GSPC Regional Threat." Washington Institute for Near East Policy. Policy Watch No. 1034. September 28, 2005. http://www.washingtoninstitute.org/templateC05.php?CID=2379.

Intergovernmental Authority on Development (IGAD). "History of IGAD." 2003. http://www.igad.org/about/index.html.

———. "IGAD Strategy." October 2003. http://www.igad.org/about/igad_strategy.pdf.

Intergovernmental Panel on Climate Change, eds. "Summary for Policymakers." In *Climate Change 2007: The Physical Science Basis.* New York: Cambridge University Press, 2007.

International Campaign to End Landmines. *Landmine Monitor Report, Guinea-Bissau, 2006.* July 1, 2006. http://www.icbl.org/lm/country/guinea_bissau/.

International Committee of the Red Cross (ICRC). "War and Water."New Press Release 99/12, March 25, 1999. http://www.icrc.org/web/eng/siteeng0.nsf/htmlall/57jpmf?opendocument.

International Crisis Group. *Crisis in Darfur*. Briefing paper on Sudan. http://www.crisisgroup.org/home/index.cfm?id=1230.

————. "Islamist Terrorism in the Sahel: Fact or Fiction." *Africa Report* 92 (March 31, 2005): 7–8, 22.

IRIN, UN Office for the Coordination of Humanitarian Affairs. "South Africa: Paying the Price for Mining." July 31, 2009. http://www.irin news.org/Report.aspx?ReportId=76780.

Israel Ministry of Foreign Affairs. "The Multilateral Negotiations." *MFA Newsletter*, November 19, 2007.

Issa, Ousseini. "Niger: Forest Squatters Demand New Homes Before Eviction." Global Information Network, New York. August 3, 2006. http://0-proquest.umi.com.usmalibrary.usma.edu.

James, Valentine U. *Environmental and Economic Dilemmas of Developing Countries: Africa in the Twenty-first Century*. Westport, CT: Praeger, 1994.

Jebb, Colonel Cindy R., Colonel Laurel J. Hummel, Lieutenant Colonel Luis Rios, and Lieutenant Colonel A. Madelfia. "Human and Environmental Security in the Sahel: A Modest Strategy for Success." In *Environmental Change and Human Security: Recognizing and Acting on Hazard Impacts*, edited by P. H. Liotta, D. A. Mouat, W. G. Kepner, and Judith M. Lancaster, 341–392. Dordrecht, The Netherlands: Springer, 2008.

Jebb, Colonel Cindy R., and Madel Abb. *Human Security and Good Governance: A Living Systems Approach to Understanding and Combating Terrorism*. Colorado Springs, CO: INSS, 2005.

Jebb, Colonel Cindy R., P. H. Liotta, Thomas Sherlock, and Ruth M. Beitler. *The Fight For Legitimacy: Democracy Versus Terrorism*. Westport, CT: Praeger Security International, 2006.

Jervis, Robert. *System Effects: Complexity in Political and Social Life*. Princeton, NJ: Princeton University Press, 1997.

Johnson, R. W. *South Africa's Brave New World: The Beloved Country Since the End of Apartheid.* London: Allen Lane, 2009.

Johnston, S., and A. Bernstein. *Voices of Anger: Protest and Conflict in Two Municipalities. Report to the Conflict and Governance Facility (CAGE).* Johannesburg: Centre for Development and Enterprise, 2007.

Johnston, N., and R. Wolmarans. "Xenophobic Violence Grips Johannesburg." *MailGuardian,* May 23, 2008. http://www.mg.co.za/articlePage.asp x?articleid=339509&area=/ breaking_news/breaking_news__national/.

Joint UNEP/OCHA Environment Unit. *Darfur Crisis, Rapid Environmental Assessment at the Kalma, Otash and Bajoum Camps.* Geneva, Switzerland, 2004.

Jones, Brian T. B. "Summary Report: Lessons Learned and Best Practices for CBNRM Policy and Legislation in Botswana, Malawi, Mozambique, Namibia, Zambia and Zimbabwe." WWF SARPO Regional Project for Community-Based Natural Resource Management (CBNRM) Capacity Building in Southern Africa. March 30, 2004. http://www.tnrf.org.

Jones, Samantha. "A Political Ecology of Wildlife Conservation in Africa." *Review of African Political Economy* 33, no. 109 (September 2006): 483–495.

Kagwanja, Peter. "Calming the Waters: The East African Community and Conflict over the Nile Resources." *Journal of Eastern African Studies* 1, no. 3 (November 2007): 321–337.

Kakonen, Jyrki, ed. *Green Security or Militarized Environment?* Aldershot, UK: Dartmouth, 1994.

Kameri-Mbote, Patricia. *Environmental Conflict and Cooperation in the African Great Lakes Region: A Case Study of the Virungas.* UNEP Environment and Conflict Prevention Initiative. Woodrow Wilson International Center for Scholars, December 2007.

————. *From Conflict to Cooperation in the Management of Trans-boundary Waters: The Nile Experience.* Heinrich Böll Foundation. 1–25. http://www.boell.org/docs/E&S_Publication_Document_FINAL. pdf.

————. *Water, Conflict, and Cooperation: Lessons from the Nile River Basin.* Navigating Peace, No. 4. Woodrow Wilson International Center for Scholars, January 2007. http://www.wilsoncenter.org/water.

Kampala Document: Towards a Conference on Security, Stability, Development and Cooperation. Document prepared by African Leadership Forum jointly with Secretariats of Organization of African Unity and the United Nations Economic Commission for Africa. New York: African Leadership Forum, 1992.

Kamudhayi, Ochieng. *Beyond the Peace Agreement: Threats and Challenges to Reconstruction and Peace in Somalia.* Nairobi: Institute for Diplomacy and International Studies, University of Nairobi. n.d.

Kansteiner, Walter H. III, and J. Stephen Morrison. *Rising U.S. Stakes in Africa: Seven Proposals to Strengthen U.S.-Africa Policy.* Report of the Africa Policy Advisory Panel of Center for Strategic and International Studies. Washington, DC: Center for Strategic and International Studies, 2004. 88–103.

Kaplan, Robert. "The Coming Anarchy." *Atlantic Monthly* 273, no. 2 (February 1994): 44–76.

Kapp, Clare. "Zimbabwe's Humanitarian Crisis Worsens." *The Lancet* 373 (2009): 447.

Kasfir, Nelson. "Sudan's Darfur: Is It Genocide?" *Current History* 104, no. 682 (May 2005): 195–202.

Katerere, Yemi, Sam Moyo, and Ryan Hill. *A Critique of Transboundary Natural Resource Management in Southern Africa.* Paper no. 1. IUCN-ROSA Series Transboundary Natural Resources Management Series (Harare, Zimbabwe: IUCN-Rosa, 2001).

Kathuri, Benson. "Egypt No Longer Has the Power to Dictate over Waters." *East African Standard*, February 23, 2005, 1.

Katunga, John. *Minerals, Forests, and Violent Conflict in the Democratic Republic of the Congo.* ECSP Report, Issue 12. Woodrow Wilson International Center for Scholars.

Kelly, C. "Summary Report Darfur Rapid Environmental Impact Assessment." CARE International/Benfield Hazard Research Centre. November 2004. http://pdf.usaid.gov/pdf_docs/PNADD448.pdf.

Kempeneers, P., E. Swinnen, and F. Fierens. *GLOBSCAR Final Report*,

AP/N7904/FF/FR-001, version 1.2. Belgium: VITO, Belgium, 2000. http://geofront.vgt.vito.be/geosuccess/documents/Final%20Report. pdf;jsessionid=668127C51913A12DB357A8569F4FE74F.

Kennedy, Lianne, and Cindy Jebb. "Non-state Actors and Transnational Issues." In *American National Security*, edited by A. Amos, A. Jordan, William J. Taylor Jr., Michael J. Meese, and S. C. Nielsen, 6th edition. Baltimore: Johns Hopkins University Press, forthcoming in 2009.

Khagram, Sanjeev, William C. Clark, and Dana Firas Raad. "From the Environment and Human Security to Sustainable Security and Development." *Journal of Human Development* 4, no. 2 (2003): 289–313.

Klare, Michael T. *Resource Wars: The New Landscape of Global Conflict.* New York: Henry Holt and Company, 2002.

Koch, Eddie. "Nature Has the Power to Heal Old Wounds: War, Peace and Changing Patterns of Conservation in Southern Africa." In *South Africa in Southern Africa: Reconfiguring the Region*, edited by D. Simon, 54–71. London: James Currey, 1998.

Kraxberger, Brennan. "The United States and Africa: Shifting Geopolitics in an 'Age of Terror." *Africa Today* 52, no. 1 (2005), http://0-proquest. umi.com.usmalibrary.usma.edu.

Kreuter, Urs P., and Randy P. Simmons. "Who Owns the Elephant? The Political Economy of Saving the African Elephant." In *Wildlife in the Marketplace*, edited by T. Anderson and P. Hill, 147–165. Lanham, MD: Rowan and Littlefield, 1995.

Kupchinsky, R. Civil Society, Water Could Become Major Catalyst for Conflict, EURASIANET.ORG. http://www.www.eurasianet.org/departments/civilsociety/articles/pp091805.shtml.

Lake, Anthony, Christine Todd Whitman, Princeton N. Lyman, and J. Stephen Morrison. "More than Humanitarianism: A Strategic U.S. Approach to Africa." Council on Foreign Relations Independent Task Force Report No. 56, 26–76. New York: Council of Foreign Relations, 2006.

Lamara, Ramtane. Permanent observer mission of the African Union to the United Nations statement by H. E. Ambassador Ramtane Lamara, commissioner for Peace and Security of the African Union, to the United Nations Security Council on the situation in Somalia, New York, December 16, 2008. http://www.africa-union.org/root/ua/Actualites/2008/dec/PSC/Statement%20by%20Amb%20%20Lamamra%20on%20Somalia%2016-12-08.pdf.

Lambert, Michael C. "Violence and the War of Words: Ethnicity v. Nationalism in the Casamance." *Africa: Journal of the International African Institute* 68, no. 4 (1998): 585–602.

Lankford, Bruce, and Thomas Beale. "Equilibrium and Non-equilibrium Theories of Sustainable Water Resources Management: Dynamic River Basin and Irrigation Behaviour in Tanzania." *Global Environmental Change* 17, no. 2 (May 2007): 168–180.

Laremont, Ricardo, and Hirach Gregorian. "Political Islam in West Africa and the Sahel." *Military Review* 86, no. 1 (January/February 2006): 27–36. http://0-proquest.umi.com.usmalibrary.usma.edu.

Lautze, Sue, and Angela Raven-Roberts. "Violence and Complex Humanitarian Emergencies: Implications for Livelihoods Models." *Disasters* 30, no. 4 (2002): 383–401.

Leader-Williams, N. *The World Trade in Rhino Horn: A Review*. Cambridge, UK: Traffic International, 1992.

Leggett, Ted, ed. *Drugs and Crime in South Africa: A Study in Three Cities*. ISS Monograph 69. Pretoria: Institute for Security Studies, 2002.

Lejano, Raul P. "Theorizing Peace Parks: Two Models of Collective Action." *Journal of Peace Research* 43, no. 5 (2006): 563–581.

Lendado, Tega. "The Green Famine of Southern Ethiopia: Myth or Real?" Ethiomedia.com, February 27, 2009. http://www.ethiomedia.com/.

Le Sage, Andre, and Nisar Majid. "The Livelihoods Gap: Responding to the Economic Dynamics of Vulnerability in Somalia." *Disasters* 26, no. 1 (2002): 10–27.

Levy, Marc A. "Is the Environment a National Security Issue?" *International Security* 20, no. 2 (Fall 1995): 35–62.

Liotta, Peter H. *The Uncertain Certainty*. Lanham, MD: Lexington Books, 2004.

Liotta, Peter H., David A. Muat, William G. Keper, and Judith M. Lancaster, eds. *Environmental Change and Human Security: Recognizing and Acting on Hazard Impacts*. Dordrecht, The Netherlands: Springer, 2007.

Liotta, Peter H., and Taylor Owen. "Sense and Symbolism: Europe Takes on Human Security." *Parameters* 36, no. 3 (2006): 85–102.

Livingstone, John K., and A. Livingstone. *A Comparative Study of Pastoralist Parliamentary Groups: Kenya Case Study*. NRI/PENHA Research Project on Pastoralist Parliamentary Groups, funded by the British Department for International Development's (DFID's) Livestock Production Programme and the CAPE Unit, African Union's Interafrican Bureau of Animal Resources (AU-IBAR). May 2005. http://www.nri.org/projects/pastoralism/kenyappgmayfinal.pdf.

Loewenberg, Samuel. "Millions in Niger Facing Food Shortages Once Again." *The Lancet* 367, no. 9521 (2006): 1474–1476.

Logistics Management Institute (LMI). *A Federal Leader's Guide to Climate Change: Policy, Adaptation, and Mitigation.* McLean, VA: LMI report, 2009.

Madavo, Callisto. *Africa: The Development Challenges of the 21st Century.* Africa Program Occasional Paper Series. Washington, DC: Woodrow Wilson International Center for Scholars, September 2005.

Malan, Mark. "AFRICOM: A Wolf in Sheep's Clothing?" Testimony to the U.S. Senate Foreign Relations Committee Subcommittee on African Affairs, August 1, 2007.

Mapping Climate Vulnerability and Poverty in Africa: Where Are the Hot Spots of Climate Change and Household Vulnerability? Report to the Department for International Development submitted by the International Livestock Research Institute (ILRI), Nairobi, Kenya, in collaboration with the Energy and Resources Institute (TERI), New Delhi, India, and the African Centre for Technology Studies (ACTS), Nairobi, Kenya, May 2006. http://www.acts.or.ke/ pubs/books/docs/ Mapping_Vuln_Africa.pdf.

Markakis, John. *Environmental Degradation and Social Conflict in the Horn of Africa.* Zurich: International Relations and Security Network, 1995. http://www.isn.ethz.ch/isn/Digital-Library/Publications/Detail/ ?ord588=grp1&ots591=0C54E3B3-1E9C-BE1E-2C24-A6A8C7060 233&lng=en&id=801.

Marks, S. A. *The Imperial Lion: Human Dimensions of Wildlife Management in Africa.* Boulder, CO: Westview, 1984.

Martin, Esmond Bradley. "Report on the Trade in Rhino Products in Eastern Asia and India." *Pachyderm* 11 (1989): 13–20.

———."The Yemeni Rhino Horn Trade." *Pachyderm 9* (1987): 13–16.

Mason, Simon A., Tobias Hagmann, Christine Bichsel, Eva Ludi, and Yacob Arsano. "Linkages Between Sub-national and International Water Conflicts: The Eastern Nile Basin." In *Facing Global Environmental Change: Environmental, Human, Energy, Food, Health and*

Water Security Concepts, edited by Hans Günter Brauch, John Grin, Czelaw Mesasz, Heinz Krummenacher, Navnita Chadha Behera, Bechir Chourou, Ursula Oswald-Spring, P. H. Liotta, and Patricia Kameri-Mbote, 325–334. Berlin: Springer-Verlag, 2007.

Mathews, Jessica T. "Redefining Security." *Foreign Affairs* 68, no. 2 (1989): 162–177.

Matowanyika, J. Z. Z. "Cast Out of Eden: Peasants Versus Wildlife Policy in Savanna Africa." *Alternatives* 16, no. 1 (1989): 30–35.

Matthew, Richard. "Environment and Security: Concepts and Definitions." In *Proceedings: Regional Asia Pacific Defence Environmental Workshop*, edited by Catherine Phinney and Kent Hughes Butts. Carlisle, PA: Center for Strategic Leadership, 1998.

McCarthy, James J., Osvaldo F. Canziani, Neil A. Leary, David J. Dokken, and Kasey S. White. *Climate Change 2001: Impacts, Adaptations and Vulnerabilities, Contribution of Working Group II to the Third Assessment Report of the Intergovernmental Panel on Climate Change*. 2001. http://www.grida.no/climate/ipcc_tar/wg2/001.htm.

McShane, Jonathan, and Thomas Adams. *The Myth of Wild Africa*. New York: Norton, 1992.

Meredith, Martin. *Diamonds, Gold and War: The Making of South Africa*. Johannesburg: Jonathan Ball, 2007.

Metcalf, Simon. "Campfire: Zimbabwe's Communal Area Management Program for Indigenous Resources." Paper prepared for the Liz Claiborne and Art Ortenburg Foundation Workshop on Community-Based Conservation, July 1993.

Metz, B., O. R. Davidson, P. R. Bosch, R. Dave, and L. A. Meyer, eds. *Climate Change 2007: Mitigation. Contribution of Working Group III to the Fourth Assessment Report of the Intergovernmental Panel on Climate Change*. Cambridge: Cambridge University Press, 2007.

Microsoft *Encarta*. "Niger River." 2006.

Miles, William F. S. "The Niger We Should Know," *Boston Globe*, August 23, 2005. http://0-proquest.umi.com.

———. "Shari'a as De-Africanization: Evidence from Hausaland." *Africa Today* 50, no. 1 (2003): 5. http://0-proquest.umi.com.usmalibrary. usma.edu.

Military Advisory Board. *National Security and the Threat of Climate Change*. Center for Naval Analysis Corporation, June 2007.

Miller, James G. *Living Systems*. New York: McGraw-Hill, 1978.

Mills, Greg, and David Williams. *Seven Battles That Shaped South Africa*. Cape Town: Tafelberg, 2006.

Mittermeier, Russell A., Cyril F. Kormos, Christina Goettsch Mittermeier, Patricio Robles Gil, Trevor Sandwith, and Charles Besancon. *Transboundary Conservation: A New Vision for Protected Areas*. Washington, DC: Conservation International, 2005.

Montague, Dana. "Stolen Goods: Coltan and Conflict in the Democratic Republic of the Congo." *SAIS Review* 22, no. 1 (2002): 103–118. http://muse.jhu.edu/journals/sais_review/v022/22.1montague.html.

Muir, Ann. *Customary Pastoral Institutions Study*. United States Agency for International Development. Nairobi, Kenya, March 2007.

Mulama, Joyce. "The Fate of the Nile in the Spotlight." Inter Press Service News Agency, March 16, 2004, 1.

Mullen, Admiral M. G. *CJCS Guidance for 2008–2009*. United States Department of Defense, 2008.

Muloongo, Keith, Roger Kibasomba, and Jemima Njeri Kariri. *The Many Faces of Human Security: Case Studies of Seven Countries in Southern Africa*. Pretoria: Institute for Security Studies, 2005.

Mutangadura, G., S. Ivens, and S. M. Donkor. "Assessing the Progress Made by Southern Africa in Implementing the MDG Target on Drinking Water and Sanitation." *Assessing Sustainable Development in Africa* (2005): 19–23. Africa's Sustainable Development Bulletin: Economic Commission for Africa, Addis Ababa.

Naicker, K., E. Cukrowska, and T. S. McCarthy. "Acid Mine Drainage from Gold Mining Activities in Johannesburg, South Africa, and Environs." *Environmental Pollution* 122 (2003): 29–40.

National Commission on Terrorist Attacks upon the United States. *The 9/11 Commission Report.* New York: W. W. Norton and Company, July 22, 2004.

National Counterterrorism Center. *2009 Counterterrorism (CT) Calendar.* McLean, VA: 2009.

National Intelligence Council. *Mapping the Global Future: Report for the National Intelligence Council's 2020 Project.* Washington, DC: NIC, December 2004.

———. "National Intelligence Estimate: The Global Infectious Disease Threat and Its Implications for the United States." *ECSPR* 6 (2000): 33–65.

National Strategy for Combating Terrorism. Washington, DC: Government Printing Office, February 2003.

National Security Strategy of the United States. Washington, DC: Government Printing Office, 1988.

———. Washington, DC: Government Printing Office, 2002.

National Water Resource Strategy (NWRS0. South African Department of Water Affairs and Forestry (DWAF): 2004. http:www.dwaf.gov.za/ Documents/Policies/NWSR/Default.htm.

NATO News. *NATO Security Science Forum on Environmental Security.* March 12, 2008. http://www.nato.int/docu/comm/2008/0803-science/e0312-summary.html.

Neethling, Theo. "Establishing AFRICOM: Pressing Questions, Political Concerns and Future Prospects." *Scientia Militaria* 36, no. 1 (2008): 31–51.

Neumann, Roderick P. *Imposing Wilderness: Struggles over Livelihood and Nature Preservation in Africa.* Berkeley: University of California Press, 1998.

Ngoma, Naison. *Prospects for a Security Community Southern Africa.* Pretoria: Institute for Security Studies, June 2005.

Nile Basin Initiative. http://www.nilebasin.org/nbibackground.htm.

North Atlantic Treaty Organization, Science for Peace and Security. "NATO-CCMS Achievements in Defence-Related Environmental Studies 1980–2001." Science for Peace and Security (SPS), North Atlantic Treaty Organization. http://www.nato.int/science/publication/coul/coul-report.htm.

Nugent, Paul. "Cyclical History in the Gambia/Casamance Borderlands: Refuge, Settlement and Islam from c. 1880 to the Present." *Journal of African History* 48 (2007): 221–243.

Nyathi, S. "No Water—No Vote." News 24, 2008. http://www.news24.com/.

Nyong, A. "Drought and Conflict in the Western Sahel: Developing Conflict Management Strategies." Event summary compiled by Alison Williams of the Woodrow Wilson International Center for Scholars, October 18, 2005.

Oberholster, P. J., and P. J. Ashton. *State of the Nation Report: An Overview of the Current Status of Water Quality and Eutrophication in South African Rivers and Reservoirs.* Pretoria: Council for Scientific and Industrial Research, 2008.

Ochoche, Obogonye, and Sunday Obogonye. "The Military and National Security in Africa." In *The Military and Militarism in Africa*, edited by E. Hutchful and A. Bathily. Dakar: CODESRIA Book Series, 1998.

Office of Dutch-Nigerian Cooperation. *Study of the Practices of Islam in Niger*. Provisional Report. April 2006.

Okoth, Godfrey P. "The New East African Community: A Defense and Security Organ as the Missing Link." In *Africa at the Crossroads: Between Regionalism and Globalization*, edited by J. Mbaku and S. Saxena, 158–159. Westport, CT: Praeger, 2004.

Pachauri, Rajendra K., and Andy Reisinger. *Climate Change 2007: Synthesis Report. Contribution of Working Groups I, II and III to the Fourth Assessment Report of the Intergovernmental Panel on Climate Change (IPCC)*. Geneva: IPCC, 2007.

Paris, Roland. "Human Security—Paradigm Shift or Hot Air?" *International Security* 26, no. 2 (2001): 87–102.

Parliamentary Debates. "South African War Mortality in the Camps of Detention. 1901, June 17." *The Parliamentary Debates, 4th series, 2nd sess. of the 27th Parliament of the United Kingdom of Great Britain and Ireland*, vol. 95. London: Wyman and Sons, 1901.

Parry, M. L., O. F. Canziani, J. P. Palutikof, P. J. van der Inden, and C. E. Hanson, eds. *Climate Change 2007: Impacts, Adaptation and Vulnerability. Contribution of Working Group II to the Fourth Assessment Report of the Intergovernmental Panel on Climate Change*. Cambridge: Cambridge University Press, 2007.

Paton, Carol. "Dam Dirty." *Financial Mail*, November 2008, 32–39.

Patterson, Kristen P. "Integrating Population, Health and Environment in Ethiopia." *Bridge: Making the Link* (2007): 1–11. Population Reference Bureau, USAID.

Paz, Reven, and Moshe Terdman. "Africa: The Gold Mine of Al Qaeda and Global Jihad." Global Research in International Affairs Center, the Project for the Research of Islamist Movements. *Occasional Papers* 4, no. 2 (June 2006).

Peace Parks Foundation. http://www.peaceparks.org.

Peluso, Nancy Lee. *Rich Forests, Poor People: Resource Control and Resistance in Java*. Berkeley: University of California Press, 1993.

Percival, V., and T. Homer-Dixon. "The Case of South Africa." In *Environmental Conflict*, edited by P. F. Diehl and N. P. Gleditsch, 13–35. Boulder, CO: Westview Press, 2001.

———. "Environmental Scarcity and Violent Conflict: The Case of South Africa." *Journal of Peace Research* 35, no. 3 (1998): 279–298.

———. *Environmental Scarcity and Violent Conflict: The Case of South Africa*. Washington, DC: American Association for the Advancement of Science, 1995.

Phillips, L. M. *With Rimington in the Boer War*. London: Edward Arnold, 1901.

Physicians for Human Rights. "Health in Ruins: A Man-Made Disaster in Zimbabwe." 2009. http://physiciansforhumanrights.org/library/report-2009-01-13.html.

Pinheiro, Isidro, Gabaake Gabaake, and Piet Heyns. "Cooperation in the Okavango River Basin: The OKACOM Perspective." In *Transboundary Rivers, Sovereignty and Development: Hydropolitical Drivers in the Okavango River Basin*, edited by A. Turton, P. Ashton, and E. Cloete, 105–118. Pretoria: African Water Issues Unit, University of Pretoria, 2003.

Pirages, Dennis. "Micro-security: Disease Organisms and Human Well-Being." *ECSPR* 2 (1995): 9–14.

Ploch, Lauren. *Africa Command: U.S. Strategic Interests and the Role of the U.S. Military in Africa*. CRS report to Congress. Washington, DC: Library of Congress, January 5, 2009, and February 2008.

Poku, Nana K., Neil Renwick, and Jaoa Gomes Porto. "Human Security and Development in Africa." *International Affairs* 83, no. 6 (2007): 1155–1171.

Polgreen, Lydia. "How Much Is Ecology to Blame for Darfur Crisis?" *New York Times*, July 22, 2007. http://www.globalpolicy.org/security/ issues/sudan/2007/0722ecology.htm.

Powell, Kristiana. *Opportunities and Challenges for Delivering on the Responsibility to Protect the African Union's Peace and Security Program*. ISS Monograph 119 (May). Pretoria: Institute for Security Studies, 2005.

Power, M. "CSIR Suspends Top Water Scientist." *The Times*, November 23, 2008. http://www.theties.co.za/PrintEdition/Insight/Article.aspx? id=890387;A.R.

Prendergast, John. *Frontline Diplomacy: Humanitarian Aid and Conflict in Africa*. Boulder, CO: Lynne Rienner, 1996.

Price-Smith, Andrew. *Chaos and Contagion: Disease, Ecology, and National Security in the Era of Globalization*. Boston: MIT Press, 2009.

Pronk, Jan. "Globalization, Poverty, and Security." In *Human and Environmental Security: An Agenda for Change*, edited by Felix Dodds and Tim Pippard. London: Earthscan, 2005.

Prunier, Gérard. *Africa's World War: Congo, the Rwandan Genocide, and the Making of A Continental Catastrophe*. Oxford: Oxford University Press, 2009.

Public Broadcasting System (PBS). "Grand Canyon." *Nature*. 2006. http://www.pbs.org/wnet/ nature/grandcanyon.

Rabasa, Angel. *Radical Islam in East Africa*. Santa Monica, CA: RAND Corporation, 2009.

RAND Corporation. *The Global Threat of New and Reemerging Infectious Diseases: Reconciling US National Security and Public Health Policy*. Santa Monica, CA: RAND, 2003.

Rejai, Mostafa, and Cynthia H. Enloe. "Nation-States and State-Nations." *International Studies Quarterly* 13, no. 2 (June 1969): 140–158.

Renner, Michael. *Fighting for Survival: Environmental Decline, Social Conflict, and the New Age of Insecurity*. New York: W.W. Norton, 1996.

Report of the High-Level Panel on Threats, Challenges, and Change to the Secretary General. http:www.un.org/secureworld/report.pdf.

Ricciuti, Edward. "The Elephant Wars." *Wildlife Conservation* 96, no. 2 (1993): 14–35.

Ross, Michael L. *Natural Resources and Civil War: An Overview with Some Policy Options*. Draft report for conference on "The Governance of Natural Resources Revenues." World Bank and Agence Francaise de Developpment. December 13, 2002. http://siteresources.world bank.org/INTCPR/1091081-1115626319273/20482496/Ross.pdf.

———. "What Do We Know About Natural Resources and Civil War?" *Journal of Peace Research* 41, no. 3 (2004): 337–356.

Rostow, W. W. *The Stages of Economic Growth: A Non-Communist Manifesto*. Cambridge: Cambridge University Press, 1960.

Rotberg, Robert I. "The New Nature of Nation-State Failure." *The Washington Quarterly* 25, no. 3 (summer 2002): 85–96.

Renner, Michael. *Fighting for Survival: Environmental Decline, Social Conflict, and the New Age of Insecurity.* New York: W. W. Norton, 1996.

Sachs, Jeffrey D. *The End of Poverty: Economic Possibilities of Our Time*. New York: Penguin Books, 2005.

———."The Geography of Economic Development." *Naval War College Review* 53, no. 4 (2000): 93–105, 101–102.

SAICE. *Presentation to Parliamentary Portfolio Committee on Water Affairs and Forestry*. Cape Town: South African Institute of Civil Engineers (SAIC), June 4, 2008.

Salih, M. A. Mohamed. Introduction in *African Pastoralism: Conflict, Institutions and Government*, edited by M. A. Mohamed Salih, Tom

Dietz, and Abdel Mohamed Ahmed, 1–22. London: Pluto Press, 2001.

Sanderson, Thomas M. "Transnational Terror and Organized Crime: Blurring the Lines." *SAIS Review* 24, no. 1 (2004): 2. http://0-proquest. umi.com.

Sas-Rolfes, M. *Rhinos: Conservation, Economics and Trade-Offs*. London: Institute of Economic Affairs, 1995.

Sayer, Jeffrey, and Bruce Campbell. *The Science of Sustainable Development: Local Livelihoods and the Global Environment*. Cambridge: Cambridge University Press, 2004.

Scharnberb, K. "Do-It-Yourself Famine Fight." *Chicago Tribune*, August 25, 2005.

Schwartz, Daniel, and Ashbindu Singh. *Environmental Conditions, Resources and Conflicts*. New York: United Nations Environmental Program (UNEP), 1999.

Seakle, K., and B. Godschalk. "Waging War over Water in Africa." *Forum* 4 (2000): 110–133.

Senge, Peter. *The Fifth Discipline: The Art & Practice of The Learning Organization*. New York: Currency Doubleday Books, 1990.

Shah, Anup. "Conflict in Africa: The Democratic Republic of Congo" *Global Issues* (2003), http://www.globalissues.org/article/87/the-democratic-republic-of-congo.

Sibanda, B. "Falling Economic Standards Cause Xenophobia in South Africa." Afrik.com, May 14, 2008. http://en.afrik.com/article13586. html.

Sivakumar, M. V. K. "Interactions Between Climate and Desertification." *Agriculture and Forest Methodology* 142, nos. 2–4 (February 2007): 143–155.

Smith, Gayle E. *In Search of Sustainable Security*. Washington, DC: Center for American Progress, June 2008.

Smith, Steve. "The Increasing Insecurity of Security Studies: Conceptualizing Security in the Last Twenty Years." *Contemporary Security Policy* 20, no. 3 (1999): 72–101.

Solomon, H., and M. van Aardt, eds. *Caring Security in Africa*. ISS Monograph 20. Pretoria: Institute for Security Studies, 1998.

Solomon, John. "The Danger of Terrorist Holes in Southern Africa." *Terrorism Monitor* 5, no. 15 (March 15, 2007), http://.jamestown.org/terrorism/news/article.php2.issue_id=4038.

South African Crime Quarterly. Pretoria, Institute for Security Studies. http://www.iss.co.za.

South African Department of Environmental Affairs and Tourism. "Overview of Integrated Environmental Management." Information Series O. 2004.

South African White Paper on Defence, 1996. Pretoria: Department of Defence, printed by 1 Military Printing Regiment, 1998.

Southern African Development Community Protocol on Shared Watercourse Systems. Gaborone: Southern African Development Community, 1995.

Spence, Mark David. *Dispossessing the Wilderness: Indian Removal and the Making of the National Parks*. Oxford: Oxford University Press, 1999.

Spenceley, Anna. "Tourism in the Great Limpopo Transfrontier Park." *Development Southern Africa* 23, no. 5 (December 2006): 649–667.

Spencer, Paul. *The Pastoral Continuum: The Marginalization of Tradition in East Africa*. Oxford: Oxford University Press, 1998.

Spies, S. B. *Methods of Barbarism: Roberts, Kitchener and Civilians in the Boer Republics January 1900–May 1902*. Cape Town: Human and Rousseau, 1977.

Spitz, R., and M. Chaskalson. *The Politics of Transition: A Hidden History of South Africa's Negotiated Settlement.* Johannesburg: Witwatersrand University Press, 2000.

Struck, Doug. "Warming Will Exacerbate Global Water Conflicts." *Washington Post*, August 20, 2007.

Sudan Tribune. "More Senegalese Troops Join Darfur Peace Mission." February 27, 2009.

Swain, Ashok. "The Nile River Basin Initiative: Too Many Cooks, Too Little Broth." *SAIS Review* 22, no. 2 (summer–fall 2002): 293–308.

Swatuk, Larry A. "Environmental Cooperation for Regional Peace and Security in Southern Africa." In *Environmental Peacemaking*, edited by K. Conca and G. Dabelko, 120–160. Washington, DC: Woodrow Wilson Center Press, 2002.

———. "Environmental Security." In *Advances in the Study of International Environmental Politics*, edited by M. Betsill, K. Hochstetler, and D. Stevis, 203–236. New York: Palgrave, 2006.

———. "Power and Water: The Coming Order in Southern Africa." In *The New Regionalism and the Future of Security and Development*, edited by B. Hettne, A. Inotai, and O. Sunkel. London: Palgrave, 2001.

Swatuk, Larry A., and Peter Vale. "Why Democracy Is Not Enough: Southern Africa and Human Security in the Twenty-first Century." *Alternatives* 24, no. 3 (1999): 361–389.

Sy, Hamadou Tidiane. "The Casamance Separatist Conflict: From Identity to the Trap of 'Identitism.'" In *Identity Matters: Ethnic and Sectarian Conflict*, edited by James L. Peacock, Patricia M. Thornton, and Patrick B. Inman, 157–170. Oxford: Berghahn Books, 2007.

Tapp, Edmund. *Manchester Museum Mummy Project: Multidisciplinary Research on Ancient Egyptian Mummified Remains*, edited by Ann Rosalie David, 99. Manchester: Manchester University Press, 1979.

Tectonidis, M. "Crisis in Niger—Outpatient Care for Severe Acute Malnutrition." *New England Journal of Medicine* 354, no. 3 (January 19, 2006), http://0-proquest.umi.com.usmalibrary.usma.edu.

Thaxton, Melissa. "Integrating Population, Health and Environment in Kenya." *Bridge: Making the Link* (2007): 1–9. Population Reference Bureau, USAID.

Theisen, Magnus Ole, and Kristian Bjarnøe Brandsegg. "The Environment and Non-state Conflicts in Sub-Saharan Africa." Paper presented at International Studies Association Convention, March 1–3, 2007.

Thomas, Louis Vincent. *Les Diola: Essai D'analyse Fonctionnelle sur une Population de Basse-Casamance.* Amsterdam: Swets and Zeitlinger, 1978.

Thurow, Roger. "Unable to Tap Power of the Nile, Ethiopia Relies on Fuel Carriers." *The Wall Street Journal*, November 26, 2003, A8.

Tignor, Robert, Jeremy Adelman, Stephen Aron, Peter Brown, Benjamin Elman, Stephen Kotkin, Xinru Liu, Suzanne Marchand, Holly Pittman, Gyan Prakash, Brent Shaw, and Michael Tsin. *Worlds Together, Worlds Apart: A History of the World from the Beginnings of Humankind to the Present*, vol. C. New York: W. W. Norton and Company, 2008.

Tilly, Charles. "War Making and State Making as Organized Crime." In *Bringing the State Back*, edited by Peter B. Evans, Dietrich Rueschmeyer, and Theda Skocpol, 169–191. Cambridge: Cambridge University Press, 1985.

Tomàs, Jordi. "La Parole de Paix N'a Jamais Tort: La Paix et la Tradition dans le Royaume d'Oussouye." *Canadian Journal of African Studies/ Revue Canadienne des Etudes Africaines* 39, no. 2 (2005): 414–441.

Tsheko, Rejoice. "Rainfall Reliability, Drought and Flood Vulnerability in Botswana." Pretoria: South African Water Research Commission. July 23, 2003. http://www.wrc.org.za.

Turton, Anthony. "Precipitation, People, Pipelines and Power: Towards a Political Ecology Discourse of Water in Southern Africa." In *Political Ecology: Science, Myth and Power*, edited by P. Stott and S. Sullivan, 132–153. London: Edward Arnold, 2000.

————. "South African Water and Mining Policy: A Study of Strategies for Transition." In *Water Transitions*, edited by D. Huitema and S. Meijerink. The Netherlands: Edgar, Elgar, 2009.

————. *Three Strategic Water Quality Challenges that Decision-Makers Need to Know About and How the CSIR Should Respond.* "A Clean South Africa," planned inaugural speech at the CSIR conference, "Science Real and Relevant," Pretoria, November 18, 2008. http://www.environment.co.za/documents/water/keynoteADDRESSCSIR 2008.pdf.

Turton, Anthony, C. Schultz, H. Buckle, M. Kgomongoe, T. Malungani, and M. Drackner. "Gold, Scorched Earth and Water: The Hydropolitics of Johannesburg." *Water Resources Development* 22, no. 2 (2006): 313–335.

Turton, Anthony, Peter Ashton, and Eugene Cloete, eds. *Transboundary Rivers, Sovereignty and Development: Hydropolitical Drivers in the Okavango River Basin.* Pretoria: University of Pretoria Centre for International Political Studies, 2003.

Tyson, P. D. "Climatic Change in Southern Africa: Past and Present Conditions and Possible Future Scenarios." *Climatic Change* 18 (1991): 241–258.

Ullman, Richard. "Redefining Security." *International Security* 8 (1983): 129–153.

United Nations. "Environmental Security: United Nations Doctrine for Managing Environmental Issues in Military Actions." http://www.acunu.org/millennium/es-un.html.

————. "Executive Summary; A More Secure World: Our Shared Responsibility." *Report of the Secretary-General's High Level Panel on Threats, Challenges and Change.* 2004. http://www.un.org/secureworld/.

————. http://www.un.org/Depts/dpko/dpko/currentops.shtml#africa.

————. http:www.un.org/genifo/bp/enviro.html.

United Nations AIDS. "Status of the Global HIV Epidemic, Report on the Global AIDS Epidemic." 2008. 30–62. http://data.unaids.org/pub/GlobalReport/2008/jc1510_2008_global_report_pp29_62_en.pdf.

United Nations Department of Peacekeeping Operations. *Environmental Guidelines for UN Field Operations.* Draft document. March 2008; final review is due June 30, 2009.

United Nations Development Programme (UNDP). *Arab Human Development Report 2002: Creating Opportunities for Future Generations.* New York: UNDP, 2002.

————. http://hdr.undp.org.

————. *Human Development Report, 1993.* http://www.undp.org/hdro/93.htm.

————. *Human Development Report, 1994.* http://www.undp.org/en/reports/global/hdr1994/.

————. *Human Development Report: International Cooperation at a Crossroads: Aid, Trade and Security in an Unequal World.* 1994.

————. *Human Development Report: International Cooperation at the Crossroads: Aid, Trade and Security in an Unequal World.* 2005.

————. *New Dimensions of Human Security.* 1994. http://hdr.undp.org/en/reports/global/hdr1994/chapters/.

United Nations Environmental Program. *Environmental for Development (GEO 4).* Valletta, Malta: Progress Press, 2007.

————. "Environment for Development, Post Conflict and Development Branch." http://postconflict.unep.ch/index.php.

————. "From Conflict to Peacebuilding, Over Forty Per Cent of Intrastate Conflicts Linked to Natures Resources, Says UNEP Report." February 20, 2009. http://www.unep.org/documents.Multilingual/Default. Print.asp?Document ID=562&Article.

United Nations Environment Programme."Environment for Development." http://www.unep.org/Documents.Multilingual/Default.asp? DocumentID=43.

————."Environment for Development, Post Conflict and Development Branch." http://postconflict.unep.ch/index.php.

————. "Natural Resource Management Critical to Peacebuilding, Over Forty Per Cent of Intrastate Conflicts Linked to Natural Resources, Says UNEP Report." February 20, 2009. http://www.unep.org/documents. Multilingual/Default.Print.asp?DocumentID=562&Article.

————. *Sudan: Post-Conflict Environmental Assessment.* June 2007. http://www.unep.org/sudan/.

————. "UNEP Post-Tsunami Recovery Activities 2004–2007." 2008. http://postconflict.unep.ch/publications/PT_recovery.pdf.

United Nations Expert Group Meeting. *Natural Resources and Conflict in Africa: Transforming a Peace Liability into a Peace Asset.* Cairo, June 17–19, 2006. http://www.un.org/africa/osaa/reports/Natural%20 Resources%20and%20Conflict%20in%20Africa_%20Cairo%20Conf erence%20ReportwAnnexes%20Nov%2017.pdf.

United Nations Fund for Population Activities (UNFPA). *Global Population and Water.* 2003. http://www.unfpa.org/publications/detail.cfm? ID=68&filterListType.

————. *State of the World Population 2004.* 2004. http://www.unfpa. org/swp.pdf/summary.pdf.

United Nations High Commissioner for Refugees (UNHCR). Packet developed for Senator Joseph Biden's trip, May 31, 2005.

———. *Report of the High-Level Panel on Threats, Challenges, and Change to the Secretary General. Briefing Note on Chad*. December 6, 2004. http://www.un.org/secureworld/report.pdf.

———. *Sudan Situation Update*. February 2005.

United Nations Integrated Regional Information Networks. "Senegal: Finding Incentives for Peace in Casamance." AllAfrica.com, June 25, 2008. http://allafrica.com/stories/200806251043.html.

———. "Senegal: Neither War Nor Peace After 25 Years." December 27, 2007. http://www.irinnews.org/Report.aspx?ReportId=76015.

United Nations News Center. "Security Council Spotlights Nexus Between Natural Resources and Conflict." June 25, 2007. http://www.un.org/apps/news/printnewsAr.asp?nid=23027.

United Nations Office for the Coordination of Humanitarian Affairs and Institute of Development Studies (UK). *Is Pastoralism Still Viable in the Horn of Africa? New Perspectives from Ethiopia*. London: Institute of Development Studies, May 2006.

———. "South Africa: Paying the Price for Mining." IRIN Humanitarian News and Analysis. http://www.irinnews.org/Report.aspx?ReportId=76780.

United Nations Office on Drugs and Crime (UNODC). *Drug Trafficking As A Security Threat in West Africa*. November 2008.

United Nations Panel of Experts. *The Illegal Exploitation of Natural Resources and Other Forms of Wealth of the Democratic Republic of the Congo*. April 12, 2001. http://www.un.org/news/dh/latest/drcongo.htm.

United Nations Peacekeeping Operations. *Principles and Guidelines*. 2008. http://pbpu.unlb.org/PBPS/Library/Capstone_Doctrine_ENG.pdf.

United Nations Security Council. "Security Council Authorizes Six-Month African Union Mission in Somalia, Unanimously Adopting Resolution 1744 (2007)." February 20, 2007. http://www.un.org/News/Press/docs/2007/sc8960.doc.htm.

United States African Command. "Partnering for Security and Stability." http://www.africom.mil/fetchBinary.asp?pdfID=20080930095216.

———. "U.S. Africa Command." http://www.africom.mil/About AFRI COM.asp.

United States Agency for International Development. *A Better Way of Working: Creating Synergies in the Pastoral Zones of the Greater Horn of Africa.* Nairobi, Kenya. Regional Enhanced Livelihoods in Pastoral Areas (RELPA): RFA—Support Documents. November 2001.

———. "Household Livestock Holding and Livelihood Vulnerabilities Status Report on PRA and Head Structure." In *Libane Zone of Somali Naitonal Regional State,* edited by Dollo Ado Woredo. Addis Ababa, Ethiopia: USAID, March, 2006.

———. *Mandera-Gedo Cross-Border Conflict Mitigation Initiative Assessment Team Report.* March 2006.

———. *Status of the Global HIV Epidemic, Report on the Global AIDS Epidemic.* 2008. 30–62. http://data.unaids.org/pub/GlobalReport/2008/jc1510_2008_global_report_pp29_62_en.pdf.

———. "Sub-saharan Initiatives." http://www.usaid.gov/locations/sub-saharan_africa/initiatives/cbfp.html.

United States Agency for International Development Office of Inspector General. *Audit of USAID/Senegal's Casamance Conflict Resolution Program.* Report No. 7-685-03-003-P. May 30, 2003.

United States Agency for International Development Sub-saharan Africa. "Congo Basin Forest Partnership." http://www.usaid.gov/locations/sub-saharan_africa/initiatives/cbfp.html.

U.S. Congress. Senate. Hillary Clinton Senate confirmation hearing transcript. Washington, DC. January 13, 2009.

U. S. Department of the Army. FM3-07, *Stability Operations.* October 2008.

U.S. Department of Defense. Directive 3000.05, *Military Support for Stability, Transition, and Reconstruction (SSTR) Operations.* November 2006.

U.S. Department of Energy, Energy Information Administration. "Environment and Renewable Energy in Africa." *Energy in Africa.* http://www.eia.doe.gov/emeu/cabs/Archives/africa/chapter7.html.

U.S. Department of State. *Dispatch* 3, no.1 (January 1, 1992): 16.

———. *Senator Paul Simon Water for the Poor Act (P.L. 109-121), Report to Congress.* June 2008.

———. *U.S. Government Counterinsurency.* January 2009.

———. Bureau of African Affairs. *Background Note: Chad.* http://www.state.gov.

———. *Background Note: Niger.* April 2006. http://www.state.gov.

———. Bureau of Democracy, Human Rights, and Labor and Bureau of Intelligence and Research. *Documenting Atrocities in Darfur.* Department of State Publication 11182. September 2004.

U.S. Environmental Protection Agency. *Environmental Security: Strengthening National Security Through Environmental Protection.* 160-F-99-001. Washington, DC: USEPA, September 1999.

Uromi, Manage Goodlae, Marc J. Stern, Cheryl Margolius, Ashley G. Lanfer, and Matthew Fladeland, eds. *Transboundary Protected Areas: The Viability of Regional Conservations Strategies.* New York: Food Products Press, 2003.

Valentine, James U. *Environmental and Economic Dilemmas of Developing Countries: Africa in the Twenty-first Century.* Westport, CT: Praeger, 1994.

Van Amerom, Marloes, and Bram Buscher. "Peace Parks in Southern Africa: Bringers of an African Renaissance?" *Journal of Modern African Studies* 43, no. 2 (2005): 159–182.

Vigne, Lucy, and Esmond Bradley Martin. "Taiwan: The Greatest Threat to Africa's Rhinos." *Pachyderm* 11 (1989): 23–25.

Wade, P. W., S. Woodbourne, W. M. Morris, P. Vos, and N. W. Jarvis. *Tier 1 Risk Assessment of Selected Radionuclides in Sediments of the Mooi River Catchment.* WRC Project No. K5/1095. Pretoria: Water Research Commission, 2002.

Waleij, Annica. *Summary Report of the NATO PfP Workshop on Environmental Security Concerns Prior to and During Peace Support and/ or Crisis Management Operations.* Activity No. SWE.2879.1 in PWP. Umea, Sweden, November 25–26, 2008. http://www.envirosecurity. org/news/uploads/FOI-R-2685--SE.pdf.

Walker, P. "African Mobs Hunt Down Immigrants." *Guardian*, May 19, 2008.

Walt, Stephen. "The Renaissance of Security Studies." *International Studies Quarterly* 35, no. 2 (June, 1991): 211–239.

Walz, Terence. *Trade Between Egypt and Bilad as-Sudan: 1700–1820.* Cairo: Institut Français d'Archeologie Orientale, 1978.

Waterbury, John. *The Nile Basin: National Determinants of Collective Action.* New Haven, CT: Yale University Press, 2001.

Watson, Robert T., et al. *Summary for Policymakers—The Regional Impacts of Climate Change: An Assessment of Vulnerability.* Intergovernmental Panel on Climate Change special report. November 1997. http://www.ipcc.ch.

Watson, Robert T., Marufu C. Zinyowera, Richard H. Moss., eds. *The Regional Impact of Climate Change: An Assessment of Vulnerability.* Cambridge: Cambridge University Press, 1997.

Watts, Susan. "An Ancient Scourge: The End of Dracunculiasis in Egypt." *Social Science and Medicine* 46, no. 7 (1998): 811–819.

Webb, Patrick, and Joachim von Braun. *Famine and Food Security in Ethiopia: Lessons for Africa.* New York: Wiley, 1994.

Welsh, Frank. *A History of South Africa*. London: Harper Collins Publishers, 2000.

Weltz, Adam. "The Spy Who Came in from the Gold: Has a Leading Water Policy Expert Been Suspended For Being a Former Spook—Or For Treading on the Mining Industry's Toes?" *Noseweek* 111 (January 2009): 30–31.

Werdmüller, V. W. "The Central Rand." In *Witwatersrand Gold—100 Years*, edited by E. S. A. Antrobus, 7–47. Pretoria: The Geological Society of South Africa, 1986.

Westenskow, Rosalie. "Intel Chief: Climate Change Threatens U.S. Security." United Press International, February 18, 2009. www.UPI.com.

Western, David. "The Undetected Trade in Rhino Horn." *Pachyderm* 11 (1989): 26–28.

Whiteside, Alan, Alex de Waal, and Tsadkan Gebre-Tensae. "AIDS, Security and the Military in Africa: a Sober Appraisal." *African Affairs* 105, no. 419 (2006): 201–218.

Whitlock, Craig. "Terror Group Expands in N. Africa/Faction Backed by al-Qaida Runs Training Camps in the Region." *Houston Chronicle*, October 6, 2006.

Williams, Paul D. "Thinking about Security in Africa." *International Affairs* 83, no. 6 (2007): 1021–1038.

Williams, Rocky. "Defence in a Democracy: The South African Defence Review and the Redefinition of the Parameters of the National Defence Debate." In *Ourselves to Know: Civil-Military Relations and Defence Transformation in Southern Africa*, edited by R. Williams, G. Cawthra, and D. Abrahams, 205–223. Pretoria: Institute for Security Studies, September, 2002.

Winde, F. "Impacts of Gold-Mining Activities on Water Availability and Quality in the Wonderfonteinspruit Catchment." In *An Assessment of Current and Future Water-Pollution Risk with Application to the Moirivierloop (Wnderfonteinspruit)*, edited by H. Coetzee, 14–38. WRC Report No. K5/1214. Pretoria: Water Research Commission, 2005.

Winde, F., and I. J. Van Der Walt. "The Significance of Groundwater-Stream Interactions and Fluctuating Stream Chemistry on Waterborne Uranium Contamination of Streams—A Case Study from a Gold Mining Site in South Africa." *Journal of Hydrology* 287 (2004): 178–196.

Wisner, Benjamin, Piers M. Blaikie, Terry Cannon, and Ian Davis. *At Risk: Natural Hazards, People's Vulnerability and Disasters.* London: Routledge, 2005.

Woodrow Wilson Center. "Africa: Rebuilding War-Torn Societies." *Centerpoint.* Washington, DC, October 2008.

Woolsey, James R., and Anne Korin. "How to Break Oil's Monopoly and OPEC's Cartel." *Innovations, Technology, Governance, Globalization* 3 (fall 2008): 35–38. http://www.mitpressjournals.org/toc/itgg/3/4.

World Bank. *Conflict in Somalia: Drivers and Dynamics.* World Bank Economic and Sector Work Project Report. January 2005. http://site resources.worldbank.org/INTSOMALIA/Resources/conflictinsomalia. pdf.

———. http://www.worldbank.org/.

———. "Niger." http://www.worldbank.org/niger.

World Climate.com. http://www.worldclimate.com.

World Health Organization. "Cholera in Zimbabwe: Update." December 26, 2008. http://www.who.int/csr/don/2008_12_26/en/index.html.

———. "Cholera in Zimbabwe: Update 2." 2009. http://www.who.int.

———. "Dracunculiasis Eradication Initiative." 2009. http://www.who. int/dracunculiasis/eradication/en/.

———. "Eradicating Guinea Worm Disease." 2008. http://whqlibdoc. who.int/hq/2008/WHO_HTM_NTD_PCT_2008.1_eng.pdf.

———. http:www.who.int/en/.

World Resources Institute. http://www.wri.org/2006.

Worldwatch Institute, ed. *State of the World 2009: Into a Warming World.* New York: W. W. Norton, 2009.

Wright, R. Michael. "Alleviating Poverty and Conserving Wildlife in Africa: An 'Imperfect' Model from Zambia." *National Association for the Practice of Anthropology (NAPA) Bulletin* 15, no. 1 (1995): 19–32.

Yao, Michel. "WHO Responding to the Drought Crisis in the Horn of Africa Region." PowerPoint presentation. World Health Organization, May 17, 2006. http://www.who.int/hac/crises/international/hoafrica/ sitreps/HoA_drought_crisis_17May2006.pdf.

Yeager, Rodger, and Norman N. Miller. *Wildlife, Wild Death: Land Use and Survival in Eastern Africa.* Albany: State University of New York Press, 1986.

Yuksel-Beten, Deniz. *Summary Final Report of the Security Science Forum on Environmental Security*, NATO Headquarters, Brussels, Belgium, June 18, 2008. http:www.eapc_sps_d_2008_007.ssf report. pdf.

Zacarias, Agostinho Zacarias. "Redefining Security." In *From Cape to Congo: Southern Africa's Evolving Security Challenges*, edited by M. Baregu and C. Lansberg, 31–52. Boulder, CO: Lynne Rienner, 2003.

Zarocostas, John. "Aid Organisations Warn Zimbabwe's Cholera Crisis Is Far From Over." *British Medical Journal* 338 (2009): 693.

LIST OF CONTRIBUTORS

Madelfia A. Abb is a retired U.S. Army lieutenant colonel who served in tactical to strategic levels in military intelligence organizations. She served as the chair of the Department of Military Science at Seton Hall University and is a 2008 recipient of the Ellis Island Medal of Honor. Her research focused on the application of living systems, complexity, and chaos theories, and she has coauthored published chapters and monographs focused on a holistic understanding of human security crises in Africa.

John Ackerman is an assistant professor of national and international security studies at the U.S. Air Force's Air Command and Staff College. He is currently investigating the intricacies of a climate action plan for the Department of Defense and the challenges and opportunities for Air Force assistance to AFRICOM and Africans.

Brent C. Bankus is a retired U.S. Army lieutenant colonel with a background in armor, cavalry, and infantry branches. He currently works in the National Security Issues Group of the U.S. Army War College, with the focus areas of environmental security, peacekeeping/stability operations, and working with the United Nations.

Jennifer Bath is an assistant professor of cellular and molecular biology at Concordia College, Moorhead, Minnesota. She is the director of research at the Concordia College Global Vaccine Institute (CCGVI), with a research focus in neglected tropical helminth diseases.

Chad Briggs is a senior fellow at the Institute for Environmental Security in The Hague (Netherlands), an assistant professor of international relations and environmental risk at Lehigh University, and a senior adviser for international security affairs to the U.S. Department of Energy's Energy and Environmental Security Directorate. His work focuses on the connections between environmental conditions and international security, environmental health risks in postconflict and postdisaster regions, and security impacts of climate and environment.

Stephen F. Burgess is an associate professor at the U.S. Air War College. His recent work focuses on U.S.-Africa policy and U.S. Africa Command as well as the role of air power in Africa. He is also working on U.S.-India relations.

Kent Hughes Butts is professor and director of the National Security Issues Group, Center for Strategic Leadership, U.S. Army War College. He served as U.S. defense attaché and security assistance officer for three African countries and is coauthor of the book *The Struggle Against Extremist Ideology: Addressing the Conditions That Foster Terrorism* (http://cbnet/orgs/usacsl/Publications/SAE-TOC.pdf).

Mark Deets, USMC, History Department, U.S. Naval Academy, is a U.S. Marine major and a foreign area officer for sub-Saharan Africa in the U.S. Marine Corps. He served as the U.S. Marine and defense attaché to Senegal, the Gambia, Guinea-Bissau, and Cape Verde from 2005–2007.

Robert Feldman is a U.S. Army lieutenant colonel and an Africa analyst with the Foreign Military Studies Office. His research interests and professional publications are in the areas of terrorism, military capabilities, and HIV/AIDS.

Elizabeth Feleke is regional program manager for West Africa at the Africa Center for Strategic Studies. She liaises and works with African regional organizations and military and civilian leaders as well as U.S. bilateral missions in West Africa on regional security issues.

Dan Henk, a social anthropologist, is the director of the Air Force Culture and Language Center at the Air University and a retired army colonel. His military service included assignments in Asia and Europe and accreditation as attaché to four African countries. His research ranges from defense budgeting and arms industries to emerging military roles and missions to human and environmental security. His most recent book is entitled *The Botswana Defense Force in the Struggle for an African Environment.*

Laurel J. Hummel is a U.S. Army colonel, an academy professor, and director of the Geography Program at West Point. She is the coeditor of "Understanding Africa: A Geographical Perspective" (2009), available at the icon of the book at http://www.dean.usma.edu/departments/geo/Geog/index.htm.

Cindy R. Jebb is a U.S. Army colonel, professor, and deputy head in the Department of Social Science at the U.S. Military Academy (West Point). Her most recent book is *The Fight for Legitimacy: Democracy Versus Terrorism*, coauthored with P. H. Liotta, Thomas Sherlock, and Ruth Beitler.

Maxie McFarland is a career defense intelligence senior executive service (DISES) professional currently serving as the deputy chief of staff, intelligence for U.S. Army Training and Doctrine Command (TRADOC). As such, he supports the army as the lead proponent for the operational environment and the resulting integration for army training, experimentation, and concept development. He also oversees the Human Terrain System program, the University of Foreign Military and Cultural Studies, the Foreign Military Studies Office, and the Joint Training CIED Operations Integration Center (JTCOIC), in support of the army and sister services training and operations.

Helen E. Purkitt is professor of international relations at the U.S. Naval Academy. She is editor of *Annual Edition: World Politics:10/11* (McGraw-Hill, 2009), coauthor of *South Africa's Weapons of Mass Destruction* (Indiana University Press, 2005), and directs a funded research project that is building a semantic wiki database of the nexus of terrorist and criminal transnational illicit networks in Africa and throughout the world.

Lieutenant Colonel Luis Rios is a U.S. Air Force lieutenant colonel and assistant professor in the Department of Geography and Environmental Engineering at West Point. His most recent work is the coauthoring of a book chapter on environmental security about the Sahel region of Africa in *Understanding Africa: A Geographic Perspective* (Center for Strategic Leadership, 2009).

Anthony Turton is an affiliated professor in the Center for Environmental Management at the University of Free State. He is a special operations veteran and still serves as an instructor at the South African National Defense College on its Executive National Security Program, which is a promotion course from colonel to brigadier general offered to various military formations across the entire African continent.

INDEX